the aloe names book

Olwen M. Grace, Ronell R. Klopper, Estrela Figueiredo & Gideon F. Smith

Pretoria
2011

This series has replaced *Memoirs of the Botanical Survey of South Africa* and *Annals of the Kirstenbosch Botanic Gardens* which SANBI inherited from its predecessor organisations.

The plant genus *Strelitzia* occurs naturally in the eastern parts of southern Africa. It comprises three arborescent species, known as wild bananas, and two acaulescent species, known as crane flowers or bird-of-paradise flowers. The logo of the South African National Biodiversity Institute is based on the striking inflorescence of *Strelitzia reginae*, a native of the Eastern Cape and KwaZulu-Natal that has become a garden favourite worldwide. It symbolises the commitment of the Institute to champion the exploration, conservation, sustainable use, appreciation and enjoyment of South Africa's exceptionally rich biodiversity for all people.

TECHNICAL EDITOR:	S. Whitehead, Royal Botanic Gardens, Kew
DESIGN & LAYOUT:	E. Fouché, SANBI
COVER DESIGN:	E. Fouché, SANBI
FRONT COVER:	*Aloe khamiesensis* (flower) and *A. microstigma* (leaf) (Photographer: A.W. Klopper)
ENDPAPERS & SPINE:	*Aloe microstigma* (Photographer: A.W. Klopper)

Citing this publication

GRACE, O.M., KLOPPER, R.R., FIGUEIREDO, E. & SMITH. G.F. 2011. The aloe names book. *Strelitzia* 28. South African National Biodiversity Institute, Pretoria and the Royal Botanic Gardens, Kew.

Citing a contribution to this publication

CROUCH, N.R. 2011. Selected Zulu and other common names of aloes from South Africa and Zimbabwe. In: O.M. GRACE, R.R. KLOPPER, E. FIGUEIREDO & G.F. SMITH, The aloe names book. *Strelitzia* 28. South African National Biodiversity Institute, Pretoria and the Royal Botanic Gardens, Kew.

ISBN: 978-1-919976-64-8
Kew edition ISBN: 978-1-84246-419-9

Obtainable from: SANBI Bookshop, Private Bag X101, Pretoria, 0001 South Africa. Tel.: +27 12 843-5000. Fax: +27 12 804-3211. E-mail: bookshop@sanbi.org.za
Website: www.sanbi.org. Distributed outside Africa by the Royal Botanic Gardens, Kew, Richmond, Surrey TW9 3AB, UK. www.kew.org, www.kewbooks.com. E-mail: publishing@kew.org
Printed by Creda Communications, 21 School Street, City and Suburban, Johannesburg, 2001 South Africa.

TABLE OF CONTENTS

Aloe gariepensis (Photographer: A.W. Klopper)

AUTHORS OF THE ALOE NAMES BOOK

Dr Olwen M. Grace
Jodrell Laboratory
Royal Botanic Gardens, Kew
Surrey TW9 3AB
United Kingdom

Email: o.grace@kew.org

Mrs Ronell R. Klopper
Biosystematics Research and Biodiversity Collections Division
South African National Biodiversity Institute
Private Bag X101, Pretoria 0001
South Africa

Email: R.Klopper@sanbi.org.za

Dr Estrela Figueiredo
Department of Botany
PO Box 77000
Nelson Mandela Metropolitan University
Port Elizabeth 6031
South Africa

and

Centre for Functional Ecology
Departamento de Ciências da Vida
Universidade de Coimbra
3001-455 Coimbra
Portugal

Email: estrelafigueiredo@hotmail.com

Prof. Gideon F. Smith FCSSA, FLS
Chief Director: Biosystematics Research and Biodiversity Collections
South African National Biodiversity Institute
Private Bag X101, Pretoria 0001
South Africa

and

John Acocks Professor of Botany
Department of Plant Science
University of Pretoria
Pretoria 0002
South Africa

and

Centre for Functional Ecology
Departamento de Ciências da Vida
Universidade de Coimbra
3001-455 Coimbra
Portugal

Email: G.Smith@sanbi.org.za

CONTRIBUTORS TO THE ALOE NAMES BOOK

Prof. Jean-Bernard Castillon* and Mr Jean-Philippe Castillon** – Selected common names of aloes from Madagascar.
* 41, Rue J. Albany, Le Tampon, 97430 La Réunion, France
** Institut Universitaire de Technologie, 40 Avenue de Soweto, 97410 Saint-Pierre, La Réunion, France

Dr António Pereira Coutinho – Selected Portuguese common names of aloes.
Centre for Functional Ecology, Departamento de Ciências da Vida, Universidade de Coimbra, 3001-455 Coimbra, Portugal

Prof. Neil R. Crouch – Selected Zulu and other common names of aloes from South Africa and Zimbabwe.
Ethnobotany Unit, South African National Biodiversity Institute, PO Box 52099, Berea Road 4007, KwaZulu-Natal, South Africa / School of Chemistry, University of KwaZulu-Natal, Durban 4041, KwaZulu-Natal, South Africa

Prof. Sebsebe Demissew – Selected common names of aloes from Ethiopia and Eritrea.
National Herbarium, Science Faculty, Addis Ababa University, PO Box 3434, Addis Ababa, Ethiopia

Dr Willem A. Jankowitz – Selected common names of aloes from Namibia.
Polytechnic of Namibia, Department of Nature Conservation, Private Bag 13388, Windhoek, Namibia

Mr Michael J. Kimberley – Selected common names of aloes from Zimbabwe.
Aloe, Cactus and Succulent Society of Zimbabwe, PO Box 85, Harare, Zimbabwe

Fr Stewart S. Lane – Selected common names of aloes from Malawi.
PO Box 354, Derdepark 0035, Gauteng, South Africa

Ms Gladys Msekandiana – Selected common names of aloes from Malawi.
National Herbarium and Botanic Gardens of Malawi, PO Box 528, Zomba, Malawi

Prof. Leonard E. Newton – Selected common names of aloes from Kenya.
Department of Botany, Kenyatta University, PO Box 43844, Nairobi 00100, Kenya

Mr Solofo E. Rakotoarisoa – Selected common names of aloes from Madagascar.
Madagascar Conservation Centre, Royal Botanic Gardens, Kew, Lot II J 131 B, Ivandry, 101 Antananarivo, Madagascar

Aloe dichotoma (Photographer: A.W. Klopper)

ACKNOWLEDGEMENTS

We are grateful to Shane Pickerill and Joseph Kelly at the Royal Botanic Gardens, Kew for their help in managing the data presented here. We thank Dr Colin Walker for eludicating the meaning of some epithets. The authors wish to acknowledge the many people who have, over the years, added common names to the PRECIS database.

The name of the photographer appears in brackets after the taxon name next to the image. The following photographers are thanked for providing images to illustrate the book (alphabetical order): S.P. Bester, C.S. Björa, J.E. Burrows, J.-P. Castillon, N.R. Crouch, S. Demissew, R. de Villiers, N. Hahn, M.J. Kimberley, J. Kirkel, A.W. Klopper, M. Koekemoer, J.C. Kruger, S.S. Lane, J.J. Meyer, G. Nichols, G. Orlando, S.E. Rakotoarisoa, H.M. Steyn, E.J. van Jaarsveld, E. van Wyk and P.J.D. Winter. Images of slides by the following people, deposited in the SANBI PRE Slide Collection, were also used (alphabetical order): P.R.O. Bally, G. Condy, D.S. Hardy, P. Joffe, J.J. Lavranos, L.C. Leach, J. Onderstall, D.C.H. Plowes, W. Rauh, G.W. Reynolds and P. Schlieben.

This book is one of the outcomes of the Aloes of the World Project, which received crucial support from the Andrew Mellon Foundation.

Aloe succotrina (Photographer: A.W. Klopper)

FOREWORD

The aloe names book is the most comprehensive information source of its kind for the aloes, a group of popular succulent plants classified by botanists in the genus *Aloe*. This book is also an attempt to associate common names with scientific names. The significance of common names is often underrated, especially in the scientific literature. Whereas the scientific names of plants convey in the first instance information on scientific classification, common names are the key to unlocking traditional knowledge residing within the human cultures that conceived them. There are, however, many instances where traditional knowledge has helped to improve scientific classifications. Common names and scientific names clearly complement each other.

Although many scientific names are based on common names, in modern times common names have also been derived from scientific ones. An often overlooked advantage of common names is their greater stability over time compared to their scientific counterparts. Incidentally, the common name 'aloe' itself encapsulates much of what can be said about the study and the significance of common names. The genus name *Aloe* is derived from a common name for one or more culturally significant members of the group in an ancient Semitic language, the identity of which is shrouded in the mists of time. The original common name entered Greek, was taken up in Latin, then passed to Old French and eventually became part of English.

In nature, aloes are found in sub-Saharan Africa, the Arabian Peninsula and on Madagascar and some smaller islands off the east coast of Africa. *Aloe* is a diverse group of more than 600 species, and its distribution range covers a part of the world that is one of the most diverse in terms of human culture. It is an area where many languages are spoken; no fewer than six of the world's major linguistic families are represented, namely the Afro-Asiatic, Nilo-Saharan, Niger-Congo, Khoesaan, Austronesian and Indo-European languages. Common names from elsewhere in the world are also included in the book, a clear reflection of the global impact of this group of plants.

An outstanding manifestation of culture is the ability to learn from the experience of others, an ability that relies on communication. The mere existence of a common name for a plant is a strong indicator of potential biocultural significance. Judged from the more than 1 500 common names recorded in this book, the significance of the aloes in human culture cannot be overestimated. The preservation of traditional knowledge related to aloes depends to a large degree on the safeguarding of both the languages and the common names that are vehicles for that knowledge.

For many aloes, published information on their significance for a particular people consists of no more than a common name. Currently, there is a global acceleration in both the rate at which biocultural knowledge is being lost and the rate at which 'smaller' languages go extinct. Hence, much field work remains to be done to record the common names of aloes and the associated traditional knowledge before it is lost. Also, from a linguistic point of view, the etymology of many common names remains unexplored.

I am confident that this valuable publication will help raise awareness of the considerable importance of aloes to humans, and of the rich contribution the group has made to the vocabulary of many languages. As names are indispensible for recording, communicating and retrieving information about a particular plant, this book will not only serve as an essential reference for those interested in aloes but should also prove useful to a broad range of other users, including plant enthusiasts, natural scientists, linguists and anthropologists.

Abraham E. van Wyk
University of Pretoria
April 2010

Aloe ferox (Photographer: A.W. Klopper)

Tapping the leaf exudate of *Aloe ferox* (Photographer: E.J. van Jaarsveld)

INTRODUCTION

Aloes are among the most familiar succulent plant groups in the world. They are valued for a multitude of traditional uses, celebrated as an iconic element of the landscape, and are immensely popular among succulent plant collectors, horticulturalists and gardeners in general. The rich cultural traditions and economic importance of aloes are reflected in the many, often descriptive, names by which they are known in languages around the world.

In recent years, there has been a significant increase in interest in the common names of plants. Among the reasons for this is that names are an information-intensive component of traditional cultures. Indices of common names are nowadays one of the inevitable starting points for anyone investigating the existing or potential value of biodiversity, as a common name often unlocks information on the indigenous uses or value of plants and animals. Many naturalists are not overly concerned with the scientific names of organisms; they are much more inclined to refer to a plant by its common name, and not least because they are easier to pronounce in a mother tongue than Latin names.

Over 600 species are presently known in the genus *Aloe* L. Recognisable by their rosettes of mostly sword-shaped, spiny, leaves and tubular flowers arranged on tall, candelabra-like inflorescences, aloes vary considerably in stature from miniature to tree-like. They are found in habitats ranging from beaches through grassland, savanna and desert to cliff faces on high-altitude mountains. The evolutionary processes that gave rise to such remarkable diversity have yet to be explained fully, and continue to frustrate the classification of aloes. Aloes are native to Africa, the Arabian Peninsula, Madagascar and islands in the western Indian Ocean. A few species, such as *Aloe arborescens* Mill., have become naturalised and invasive in parts of Europe, Asia and the Americas.

Aloes attract curiosity among scientists and laypeople alike, resulting in an ever-increasing body of literature covering topics from DNA to pollinators. This book arose from the Aloes of the World Project, an initiative to compile a global repository of scholarly information about aloes, hosted by the South African National Biodiversity Institute (SANBI) and supported by the Andrew Mellon Foundation.

Useful aloes

The global popularity of aloes in horticulture dates to the sixteenth century, when they became fashionable in collections of exotic plants in Europe. However, aloes have a much longer history in medicine: the uses, properties and importance of *Aloe vera* (L.) Burm.f. have been documented for thousands of years. The uses of aloes for cosmetics, medicine and incense and as an ingredient in embalming substances are mentioned both in records throughout antiquity, including those of the ancient Egyptians, Romans and Greeks, and in religious texts such as the Bible and Qur'an. About 30 kg of an embalming mixture comprising the powdered leaves of *Aloe vera* and myrrh were taken to the garden tomb by Nicodemus after the crucifixion of Jesus (The Bible, John 19: 39).

The major commodity taken from *Aloe vera* and other aloes is a dried, crystalline preparation of the leaf exudate known collectively as 'drug aloes' or 'bitter aloes', which is used to treat digestive complaints. Drug aloes are usually named according to their species or geographic region of origin; for example, Cape aloes is made from the leaf exudate of the South African species *Aloe ferox* Mill. Safety concerns have led to a decline in the use of drug aloes in Europe and the

United States in recent years. However, there is a growing demand for the colourless mesophyll tissue of the leaves ('aloe gel') of *Aloe vera* and certain other species. The dried, powdered tissue is used in foods, cosmetics and household commodities the world over. People have depended on aloes for livelihood security for centuries. While *Aloe vera*, *Aloe ferox* and certain other *Aloe* species are grown as crops for the natural products from their leaves, natural populations of these and other species are wild harvested, in certain cases justifying serious concerns for the sustainability and quality of *Aloe* derivatives.

Conservation and trade

Aloes occur naturally in places of exceptional biodiversity. The aloes of Madagascar, for instance, are all endemic: they are not found naturally anywhere else on Earth. The majority of aloes are adapted to grow in specific conditions in a single, localised area; relatively few are found over an extensive range—*Aloe arborescens*, *Aloe maculata* All. and *Aloe myriacantha* (Haw.) Schult. & Schult.f. being notable exceptions. Species growing in arid habitats are likely to become increasingly important in regions experiencing less rainfall and raised temperatures caused by human-induced climate change. In addition to habitat loss, the greatest threats to aloes in the wild are plant collectors and people harvesting the leaves and other plant parts for natural products. Accordingly, all species except *Aloe vera* are protected by the Convention on the International Trade in Endangered Species of Wild Fauna and Flora (CITES) and appropriate licenses are required for their transportation across international borders. Permits issued in advance by local conservation authorities are, likewise, required in most countries for those wishing to collect and study aloes.

Plant names

The science of naming and classifying living organisms is called taxonomy. When a plant is described for the first time, it is given a Latin name comprising a unique combination of a genus name (e.g. *Aloe*) and specific epithet (e.g. *murina*), which together make a species name. The taxonomist who describes a new species is known as the author of that plant name. Written in full, a species name includes the author's name or a standard abbreviation, such as *Aloe murina* L.E.Newton, for Professor Leonard Newton who first described the plant in 1994. Two or more forms of a species may be further distinguished by an additional Latin epithet at the ranks of subspecies (abbreviated to subsp.) or variety (var.). Latin plant names are usually very descriptive and their etymologies (origin and history) often provide insights into the circumstances in which the plant was first described. Plant names may change as scientists make new discoveries or uncover previous errors. Consequently, many plants have names, called synonyms, by which they are no longer known. These are not necessarily incorrect; they could simply be alternative names under which the plant was, or could be, known. The *International Code of Botanical Nomenclature* ensures that plant names are used correctly.

Although all plant species ideally have a single accepted Latin name, they may also be known in other languages by vernacular, or common, names. These arise as people develop the need to classify and communicate information about a plant, often describing its cultural importance or socio-economic value. Common names are, therefore, very useful for understanding the local importance of plants. Species known by many common names, such as *Aloe vera*, are often widespread and of great importance to people.

Whereas a Latin plant name is applied to a single species, common names may be applied to many, sometimes unrelated, species. In Malawi, for instance, two species of *Aloe* are recognised as the male and female forms of one common name (Msekandiana 2009, personal communication). The potential ambiguity of common names is evident in the literature. For example, in the Old Testament of the Bible (Numbers 24: 6, Psalm 45: 9 and Canticles 4: 14), 'aloes' or 'lign aloe' refer to *Aquilaria malaccensis* Lam., an unrelated species in the plant family Thymelaeaceae. 'Aloes' in the New Testament (John 19: 39), however, refers to a product derived from a true species of aloe. It has been claimed that it refers to the leaf extract of a species of aloe that originated from the island of Socotra. If this is correct, the species in question is most likely *Aloe perryi* Baker (*Aloe succotrina* Weston has also been proposed but seems unlikely as this species is found only in the extreme southwestern Cape of South Africa and does not occur on Socotra). Latin plant names are helpful in reducing ambiguity and allowing information to be exchanged in any language, because each plant usually has only one scientific name.

About this book

This book is an annotated list of names for the genus *Aloe*. It is intended to provide a quick reference for checking names and key information about aloes. Accepted names are arranged alphabetically in Part I, or can be determined from lists of synonyms, common names and products in Part IV.

There are 624 accepted Latin names, including 119 subspecies or varieties and 26 natural hybrid aloes, and 490 synonyms in Part I. The exact application is not known for 28 names with 11 synonyms (listed in Part II). Synonyms in genera largely subsumed under *Aloe*, such as *Leptaloe*, *Lomatophyllum* and *Guillauminia*, are reflected in the synonymy provided, but misapplied and invalid names are excluded. Accepted Latin names are arranged alphabetically. Author abbreviations follow the International Plant Names Index (www.ipni.org). The meanings of some abbreviations of taxonomic terms in this book are given in the glossary below. For each aloe in this book, information is arranged under the subheadings 'Synonyms', 'Etymology', 'Common names' and 'Products'. Subheadings are not shown if information has not been documented.

Etymologies describing the origin and history of a plant name are provided for some 881 Latin epithets. A total of 1 519 common names of aloes in 164 languages from around the world are compiled here for the first time. A numeric reference in superscript is given for each common name; the corresponding literature is listed in the References (Part III). Common names have not been edited according to conventions accepted in the various languages to avoid introducing errors, as the authors' expertise covers a small proportion of the languages included in this book. For this reason, common names may be repeated with different spellings. The widely used common names, such as 'aloe' and 'aalwyn' (in Afrikaans), are listed under *Aloe* L. because these names are used interchangeably for all species of *Aloe*. Languages follow the Ethnologue language compendium (www.ethnologue.org). Standardised natural products prepared from aloes are listed as 'Products', with notes on their composition as stipulated in regional pharmacopoeias and other industry standards. These are by no means an exhaustive list of products containing aloes, but are the ones for which standards are available. Products and the source species from which they are derived can be looked up in Part IV.

The data in this book were current at the time of publication. While every effort has been made to ensure accuracy, the authors would be grateful for notification of errors or omissions.

Glossary

nom. illegit.: illegitimate name; *nomen illegitimum*. A validly published name that is not in accordance with one or more rules of the *International Code of Botanical Nomenclature*, principally those on superfluous names and homonyms.

nom. superfl.: superfluous name; *nomen superfluum*. A name applied to a taxon circumscribed by the author to definitely include the type of a name which ought to have been adopted, or of which the epithet ought to have been adopted under the rules of the *International Code of Botanical Nomenclature*.

p.p.: in part; *pro parte*. A polymorphic taxon is sometimes subsequently split into several taxa. The name of the original taxon is then a *pro parte* synonym of the two or more new taxon names, because part of the original group of plants is now in one taxon, part in another.

Bibliography

ADMIRAAL, J. 1984. *Die Psalmodie van die plante. Plante in en rondom die Bybel*. Mediese Universiteit van Suid-Afrika, Medunsa.

BISBY, F.A. 1994. *Plant names in botanical databases*. Plant Taxonomic Database Standards No. 3. Hunt Institute for Botanical Documentation, Pittsburg, California.

EGGLI, U. & NEWTON, L.E. 2004. *Etymological dictionary of succulent plant names*. Springer-Verlag, Berlin, Heidelberg.

FAROOQI, M.I.H. 2000. *Plants of the Qu'ran*. 5th Ed. Sidrah, Lucknow, Uttar Pradesh.

GRACE, O.M., SIMMONDS, M.S.J., SMITH, G.F. & VAN WYK, A.E. 2008. Documented utility and biocultural value of *Aloe* L. (Asphodelaceae): a review. *Economic Botany* 63: 167–178.

GUGLIELMONE, L., GALLO, L., MEREGALLI, M., SMITH, G.F. & FIGUEIREDO, E. 2009. Allioni's *Aloe* names (Asphodelaceae): nomenclature and typification. *Bothalia* 39: 177–183.

HEPPER, F.N. 1981. *Bible plants at Kew*. 2nd impression with revisions. Royal Botanic Gardens, Kew.

MCNEILL, J., BARRIE, F.R., BURDET, H.M., DEMOULIN, V., HAWKSWORTH, D.L., MARHOLD, K., NICOLSON, D.H., PRADO, J., SILVA, P.C., SKOG, J.E., WIERSEMA, J.H. & TURLAND, N.J. (2006). *International Code of Botanical Nomenclature (Vienna Code)* adopted by the Seventeenth International Botanical Congress, Vienna, Austria, July 2005. *Regnum Vegetabile* 146: 1–568. Gantner Verlag, Liechtenstein.

MUSSELMAN, L.J. 2007. *Figs, dates, laurels and myrrh. Plants of the Bible and the Quran.* Timber Press, Portland, Oregon.

PATERSON, J. & PATERSON, K. 1986. *Consider the lilies: plants of the Bible*. Thomas Y. Crowell, New York.

SMITH, G.F. 1993. Familial orthography: Aloeaceae vs. Aloaceae. *Taxon* 42: 87–90.

SMITH, G.F. & VAN WYK, B-E. 1996. *Aloe succotrina* and Reynolds' book on the aloes of South Africa. *Aloe* 33: 57–58.

WALKER, W. 1957. *All the plants of the Bible*. Harper & Brothers, New York.

Aloe dichotoma (Photographer: H.M. Steyn)

PART I: ACCEPTED NAMES WITH SYNONYMS, ETYMOLOGY, COMMON NAMES AND PRODUCTS

Aloe L.

Etymology

Aloe: Uncertain; two possibilities have been suggested: from Arabic 'alloch' or 'alloeh', a vernacular name for members of the genus used medicinally; or from Greek 'aloë', the dried juice of aloe leaves, this word is akin to or derived from earlier Semitic ('alloeh'), Hebrew ('ahalim' or 'allal', i.e. bitter) and Sanskrit words.

Common names

aalewee [108] [Afrikaans; Dutch]
aalwee [108] [Afrikaans]
aalwyn [108] [Afrikaans]
aloe [108] [English]
aloë [108] [Dutch]
aloes [119, 122] [English; Dutch]
äray [40] [Tigrigna]
argeesaa [40] [Oromo, West Central]
chidzima mliro [82] [Ngoni]
chigiakia [8] [Tswa]
chikowa [8, 103] [Shona]
chintembwe [68] [Nyanja]
chinthembwe [82] [Nyanja] [82] [Tumbuka]
chinungu [8] [Shona]
chinyangami [8] [Tonga]
chitembwe [68] [Nyanja]
chiwiriwiri [68] [Tumbuka]
daar [40] [Somali]
gabar [40] [Somali]
gavakava [7, 8] [Shona]
gave wamtchanga [82] [Sena]
godole uta [40] [Wolaytta]
godzongo [8] [Shona]
gweravana [8] [Shona]
hlaba [58] [Sotho, Southern]
icena [7, 8] [Ndebele]
`iegedel fuga [40] [Guragigna]
imboma [46] [Zulu]
iwani [82] [Nyika]
kakaruamba [5] [Zinza]
kakruamba [5] [Zinza]
kanembe [103] [Tonga]
khonje wa mtchire [82] [Lomwe]
khuzi [82] [Ngoni]
lekhala [58] [Sotho, Southern]
lichongwe [82] [Sena] [68] [Yao]
losa [5] [Shambala] [5] [Sukuma]
lugaka [5] [Sukuma]
mala [54] [Umbundu]
manasvato [29] [Unspecified language]
manyesa [82] [Lomwe]
matiso [40] [Tigrigna]
mdyang'oma [82] [Ngoni]
mgaka [5] [Sukuma]
mhangani [8] [Tswa]
mkakruamba [5] [Zinza]
namanyesa [82] [Nyanja]
nemba [82] [Ngoni]
ngaka [5] [Sukuma]
ober [40] [Tigrigna]
're [40] [Saho] [40] [Tigrigna]
`riet [40] [Amharic]
rukaka [5] [Zinza]
rumhangamhuno [8] [Shona]
ruvati [8] [Shona]
senjela [82] [Lomwe] [82] [Nyanja] [82] [Yao]
senjerere [82] [Lomwe]
zabur [40] [Tigrigna]

General aloe-derived products

Aloes: The residue obtained by evaporating the juice of the leaves of various species of *Aloe*[89].

Drug aloes: Dried latex or bitter yellow leaf exudate from leaf bundle sheath cells, used as a cathartic[87].

Powdered aloes: The powdered residue obtained by evaporating the juice of the leaves of various species[21].

Zanzibar aloes: Prepared from the juice of leaves[92].

Aloe aageodonta L.E.Newton

Etymology

aageodonta: With hard teeth, from Greek 'aages' (hard), 'odous, odontus' (teeth).

Aloe abyssicola Lavranos & Bilaidi

Etymology

abyssicola: Living in abysses, from Latin 'abyssus' (abyss), '-cola' (-dwelling).

Aloe aculeata Pole-Evans

Aloe aculeata (Photographer: M.J. Kimberley)

Etymology

aculeata: For the spiny leaf surface, from Latin 'aculeatus' (prickly, pointed).

Common names

knoppiesaalwyn [55, 109] [Afrikaans]
ngopane [55] [Sotho, Southern]
ngopani [53, 55, 123] [Sotho, Northern]
red hot poker aloe [55, 126] [English]
sekope [53, 123] [Sotho, Northern]
white-thorn aloe [55, 109] [English]
witdoringaalwyn [55, 109] [Afrikaans]

Aloe acutissima H.Perrier subsp. *acutissima* var. *acutissima*

Etymology

acutissima: For the leaves, from Latin 'acutus' (acute), superlative.

Aloe acutissima H.Perrier subsp. *acutissima* var. *antanimorensis* Reynolds

Etymology

acutissima: For the leaves, from Latin 'acutus' (acute), superlative.
antanimorensis: For the occurrence at Antanimora, in Madagascar.

Aloe acutissima H.Perrier subsp. *acutissima* var. *fiherenensis* J.-B.Castillon

Etymology

acutissima: For the leaves, from Latin 'acutus' (acute), superlative.
fiherenensis: For the occurrence along Fiherenana, in Madagascar.

Aloe acutissima H.Perrier subsp. *acutissima* var. *isaloana* J.-B.Castillon

Etymology

acutissima: For the leaves, from Latin 'acutus' (acute), superlative.
isaloana: For the occurrence near the Massif de l'Isalo, in Madagascar.

Aloe acutissima H.Perrier subsp. *itampolensis* Rebmann

Synonyms

A. acutissima H.Perrier var. *itampoloana* J.-B.Castillon

Etymology

acutissima: For the leaves, from Latin 'acutus' (acute), superlative.
itampolensis/*itampoloana*: For the occurrence near Itampolo, in Madagascar.

Aloe adigratana Reynolds

Synonyms

A. eru A.Berger var. *hookeri* A.Berger

Etymology

adigratana: For the occurrence at Adigrat, in Ethiopia.
eru: For the common name of the plant, 'eru', in Ethiopia.
hookeri: For Sir Joseph Dalton Hooker (1817–1911), British botanist and explorer, and director of the Royal Botanic Gardens, Kew.

Common names

iret [55, 102] [Tigrigna]

Aloe affinis A.Berger

Aloe affinis (Photographer: SANBI, P. Joffe)

Synonyms

A. immaculata Pillans

Etymology

affinis: For the relationship to *Aloe zebrina*, although it is no closer to that species than to any other maculate aloe[53, 114], from Latin 'affinis' (allied to).
immaculata: For the unspotted leaves, from Latin 'im-' (not), 'maculatus' (maculate).

Common names

bontaalwyn [55, 109] [Afrikaans]
inhlaba [55] [Swati]
spotted aloe [55, 109] [English]

Aloe africana Mill.

Aloe africana (Photographer: E.J. van Jaarsveld)

Synonyms

A. africana Mill. var. *angustior* Haw.
A. africana Mill. var. *latifolia* Haw.
A. angustifolia Haw.
A. bolusii Baker

Aloe acutissima subsp. *itampolensis*

A. perfoliata L. var. *β* L.
A. perfoliata L. var. [*β*] *africana* (Mill.) Aiton
A. pseudoafricana Salm-Dyck
Pachidendron africanum (Mill.) Haw.
P. africanum (Mill.) Haw. var. *angustum* Haw.
P. africanum (Mill.) Haw. var. *latum* Haw.
P. angustifolium (Haw.) Haw.

Etymology

africana/africanum: For the occurrence in Africa.
angustifolia/angustifolium: For the narrow leaves, from Latin 'angustus' (narrow), '-folius' (leaved).
angustior: For the narrower leaves, from Latin 'angustus' (narrow), comparative.
angustum: For the narrow leaves, from Latin 'angustus' (narrow).
bolusii: For Harry Bolus (1834–1911), English-born South African stockbroker and botanist, who collected the type.
latifolia: For the wide leaves, from Latin 'latus' (broad), '-folius' (leaved).
latum: For the wide leaves, from Latin 'latus' (broad).
perfoliata: For the stem passing through the leaves, i.e. the leaves are amplexicaul, from Latin 'per' (through), 'folia' (leaf).
pseudoafricana: For resembling *Aloe africana*, from Greek 'pseudo-' (false).

Common names

Uitenhaags-aalwee [108] [Afrikaans]
Uitenhaagsaalwyn [55, 109, 123, 124] [Afrikaans]
Uitenhaagse aalwyn [53] [Afrikaans]
Uitenhaagse-aalwee [99] [Afrikaans]
Uitenhaagseaalwyn [62] [Afrikaans]
Uitenhaagse-aalwyn [99] [Afrikaans]
Uitenhage aloe [55, 99, 109, 123, 124] [English]

Aloe ahmarensis Favell, M.B.Mill. & Al-Gifri

Etymology

ahmarensis: For the occurrence at Al Ahmar, in Yemen.

Aloe albida (Stapf) Reynolds

Aloe albida (Photographer: J.E. Burrows)

Synonyms

A. kraussii Schönland (nom. illegit.)
A. kraussii Baker var. *minor* Baker
A. myriacantha (Haw.) Schult. & Schult.f. var. *minor* (Baker) A.Berger
Leptaloe albida Stapf

Etymology

albida: For the flower colour, whitish, from Latin 'albus' (white).
kraussii: For Dr Ferdinand F. von Krauss (1812–1890), German scientist, director of the Stuttgart Natural History Museum, traveller and collector in South Africa.
minor: For the smaller size, from Latin, comparative of 'parvus' (small).
myriacantha: For the many fine teeth on the leaf margins, although the spines are no more numerous than those of other grass aloes[123], from Greek 'myrios' (numerous), 'akantha' (thorn, spine).

Aloe albiflora Guillaumin

Synonyms

Guillauminia albiflora (Guillaumin) A.Bertrand

Etymology

albiflora: White-flowered, from Latin 'albus' (white), 'florus' (flowered).

Aloe albiflora (Photographer: SANBI)

Aloe albostriata T.A.McCoy, Rakouth & Lavranos

Etymology

albostriata: With white stripes, from Latin 'albus' (white), 'striatus' (striped).

Aloe albovestita S.Carter & Brandham

Aloe albovestita (Photographer: G. Orlando)

Etymology

albovestita: For the heavy bloom on the tepals, from Latin 'albus' (white), 'vestitus' (clothed).

Aloe aldabrensis (Marais) L.E.Newton & G.D.Rowley

Synonyms

Lomatophyllum aldabrense Marais

Etymology

aldabrense/aldabrensis: For the occurrence on Albabra Island.

Common names

zanana mowo [44] [Unspecified language]

Aloe alexandrei Ellert

Etymology

alexandrei: For Mr Alexandre Viossat, who discovered the plant.

Aloe alfredii Rauh

Etymology

alfredii: For Alfred Razafindratsira (fl.1941–1987), Madagascan plant collector and owner of a plant nursery.

Aloe alooides (Bolus) Druten

Synonyms

A. recurvifolia Groenew.
Notosceptrum alooides (Bolus) Benth.
Urginea alooides Bolus

Etymology

alooides: For resembling an aloe, as it was first described and named in the genus *Urginia*, from Greek '-oides' (resembling).

Aloe alooides (Photographer: G.F. Smith)

recurvifolia: For the curved leaves, from Latin 'recurvus' (curved backwards), '-folius' (leaved).

Common names

eel aloe [55, 99] [English]
Graskop aloe [123, 124] [English]
Graskopaalwyn [123, 124] [Afrikaans]
rokaalwyn [55, 109] [Afrikaans]
skirt aloe [55, 109] [English]

Aloe ambigens Chiov.

Etymology

ambigens: Application obscure, from Latin 'ambigere' (to doubt).

Aloe ambositrae J.-P.Castillon

Etymology

ambositrae: For the occurrence at Ambositra, in Madagascar.

Aloe ambrensis J.-B.Castillon

Etymology

ambrensis: For the type locality, at Cap d'Ambre, in Madagascar.

Aloe amicorum L.E.Newton

Etymology

amicorum: For the friends of the Mountain Club of Kenya expedition on which the taxon was discovered, from Latin 'amicus' (friend).

Aloe ammophila Reynolds

Etymology

ammophila: For the preferred sandy habitat, from Greek 'ammos' (sand), '-philos' (friend).

Aloe ampefyana J.-B.Castillon

Etymology

ampefyana: For the occurrence at Ampefy, in Madagascar.

Aloe amudatensis Reynolds

Etymology

amudatensis: For the occurrence at Amudat, in Uganda.

Aloe alooides

Aloe andongensis Baker var. *andongensis*

Aloe andongensis var. *andongensis* (Photographer: G. Nichols)

Etymology

andongensis: For the occurrence at Pungo Andongo, in Angola.

Aloe andongensis Baker var. *repens* L.C.Leach

Aloe andongensis var. *repens* (Photographer: SANBI, P. Joffe)

Etymology

andongensis: For the occurrence at Pungo Andongo, in Angola.
repens: For the prostrate habit, from Latin 'repens' (creeping).

Aloe andringitrensis H.Perrier

Aloe andringitrensis (Photographer: SANBI, G.W. Reynolds)

Etymology

andringitrensis: For the occurrence on the Andringitra Mountains, in Madagascar.

Aloe angelica Pole-Evans

Etymology

angelica: For Mrs (Angelique) R.C. Wallace, whose husband was a former chief engi-

Aloe angelica (Photographer: G.F. Smith)

neer of the South African Railways and Harbours (SARAH), and who brought the plant to the attention of Dr I.B. Pole-Evans.

Common names

Wylliespoort aloe [55, 57, 123, 124] [English]
Wylliespoortaalwyn [9, 55, 57, 123, 124] [Afrikaans]

Aloe angolensis Baker

Etymology

angolensis: For the occurrence in Angola.

Common names

nllhoq'uru [55] [Nama]
nllhoq'ùrù [72] [Ju|'hoan]

Aloe anivoranoensis (Rauh & Hebding) L.E.Newton & G.D.Rowley

Synonyms

Lomatophyllum anivoranoense Rauh & Hebding

Etymology

anivoranoense/anivoranoensis: For the occurence near Anivorano, in Madagascar.

Aloe ankoberensis M.G.Gilbert & Sebsebe

Aloe ankoberensis (Photographer: G. Orlando)

Etymology

ankoberensis: For the occurrence at Ankober, in Ethiopia.

Common names

merarie [40] [Amharic]

Aloe antandroi (Decary) H.Perrier subsp. *antandroi*

Synonyms

Gasteria antandroi Decary

Etymology

antandroi: For the occurrence at Antandroi in Magadascar or in the territory of the Antandroi tribe.

Aloe antandroi subsp. *antandroi* (Photographer: S.E. Rakotoarisoa)

Common names

tsikivahonbaho [55, 91] [Unspecified language]
tsikyvahombaho [102] [Malagasy]

Aloe antandroi (Decary) H.Perrier subsp. *toliarana* J.-B.Castillon

Etymology

antandroi: For the occurrence at Antandroi in Magadascar or in the territory of the Antandroi tribe.
toliarana: For the occurrence near Toliara, in Madagascar.

Aloe antonii J.-B.Castillon

Etymology

antonii: For Antoine Castillon, grandson of the author.

Aloe antsingyensis (Leandri) L.E.Newton & G.D.Rowley

Synonyms

Lomatophyllum antsingyense Leandri

Etymology

antsingyense/antsingyensis: For the occurrence at Antsingy, in Madagascar.

Aloe arborescens Mill. subsp. *arborescens*

Aloe arborescens subsp. *arborescens* (Photographer: N.R. Crouch)

Synonyms

A. arborea Medik.
A. arborescens Mill. var. *frutescens* (Salm-Dyck) Link
A. arborescens Mill. var. *milleri* A.Berger
A. arborescens Mill. var. *natalensis* (J.M.Wood & M.S.Evans) A.Berger
A. arborescens Mill. var. *pachythyrsa* A.Berger
A. arborescens Mill. var. *viridifolia* A.Berger
A. frutescens Salm-Dyck
A. fruticosa Lam.
A. fulgens Tod.
A. natalensis J.M.Wood & M.S.Evans
A. perfoliata L. var. [α] *arborescens* (Mill.) Aiton
A. perfoliata L. var. η L.
A. sigmoidea Baker (possibly a hybrid)
Catevala arborescens (Mill.) Medik.

Etymology

arborea: For being tree-like, from Latin 'arbor' (tree).

arborescens: For becoming tree-like, from Latin 'arbor' (tree), although the plant is not a tree but a large, much-branched shrub[53].

frutescens: For becoming shrubby, from Latin 'frutex' (shrub).

fruticosa: For being shrubby, from Latin 'frutex' (shrub).

fulgens: Probably for the flowers, from Latin 'fulgens' (shining, bright-coloured).

milleri: For Philip Miller (1691–1771), British botanist and horticulturalist at the Chelsea Physic Garden, who published the eighth edition of his famous Gardener's Dictionary in 1768.

natalensis: For the occurrence in the province of Natal (KwaZulu-Natal), in South Africa.

pachythyrsa: For the dense inflorescence, from Greek 'pachys' (thick) and Latin 'thyrsus' (thyrse).

perfoliata: For the stem passing through the leaves, i.e. the leaves are amplexicaul, from Latin 'per' (through), 'folia' (leaf).

sigmoidea: Application obscure, meaning curved like an 's', from Greek 'sigma' (the letter 's').

viridifolia: For the green leaves, from Latin 'viridis' (green), '-folia' (leaved).

Common names

aloé [43, 99, 104] [Portuguese] [30, 73] [Spanish]
aloe a candelabro [97] [Italian]
aloé arborescente [30, 73] [Spanish]
aloé candelabro [4] [Portuguese]
aloé-candelabro [43] [Portuguese]
aloés [104] [Portuguese]
aloès arborescent [10] [French]
amaposo [55] [Lozi]
atzavara [73] [Spanish]
azebra [99] [Portuguese]
babosa [43, 99, 104] [Portuguese]
balsemera [73] [Spanish]
belarmintz [73] [Spanish]
candelabra aloe [4, 10, 56, 97] [English]
candelabra plant [97] [English]
candelabra-plant [56] [English]
chintembwe [55, 102] [Nyanja]
chitseyse [37] [Unspecified language]
chiwiriwiri [55, 102] [Nyanja]
erva-babosa [99] [Portuguese]
foguetes de Natal [99] [Portuguese]
foguetes-de-Natal [43] [Portuguese]
ikhala [94, 96, 99] [Xhosa]
inhlaba [55] [Swati] [65] [Zulu]
inhlaba-encane [55, 94, 96, 99] [Zulu]
inhlazi [49, 53, 55, 99] [Zulu]
inkalame [65] [Zulu]
inkalane [50, 55, 122, 123] [Zulu]
inkalane encane [53, 55] [Zulu]
inkalane-encane [94, 96, 99] [Zulu]
inkalene encane [99] [Zulu]
iposo [55] [Bemba]
Kidachi aloe [10] [English]
Kidachi Japanese aloe [55] [English]
krans aloe [99] [English]
kransaalwyn [9, 55, 57, 65, 81, 83, 94, 96, 99, 108, 109, 122, 123, 124] [Afrikaans]
krantz aloe [10, 55, 57, 64, 94, 96, 99, 109, 122, 123, 124] [English]
mountain bush aloe [4, 10, 55] [English]
octopus plant [10, 97] [English]
octopus-plant [56] [English]
Oldenland's bush aloe [55, 65, 126] [English]
pulpos [30, 73] [Spanish]
sabila [55] [English]
sayyan [53, 99] [Sotho, Northern]
sword aloe [99] [English]
torch plant [10, 97] [English]
torchplant [56] [English]
tshikhopha [57] [Venda]
umhlabana [55, 94, 96, 99] [Zulu]
unomaweni [94, 96, 99] [Xhosa]
vela [104] [Portuguese]
woody aloe [97] [English]
xitretre [4] [Ronga]
zabila [30, 73] [Spanish]
zebra [99] [Portuguese]

Aloe arborescens Mill. subsp. *mzimnyati* van Jaarsv. & A.E.van Wyk

Etymology

arborescens: For becoming tree-like, from Latin 'arbor' (tree), although the plant is not a tree but a large, much-branched shrub[53].

mzimnyati: For the occurrence along the lower Mzimnyati River in KwaZulu-Natal, South Africa.

Aloe archeri Lavranos subsp. *archeri*

Etymology

archeri: For Philip G. Archer (1922–), British accountant and succulent plant enthusiast resident in Kenya 1950–1974.

Aloe archeri Lavranos subsp. *tugenensis* (L.E.Newton & Lavranos) Wabuyele

Synonyms

A. tugenensis L.E.Newton & Lavranos

Etymology

archeri: For Philip G. Archer (1922–), British accountant and succulent plant enthusiast resident in Kenya 1950–1974.

tugenensis: For the occurrence in the Tugen Hills, in Kenya.

Aloe arenicola Reynolds

Aloe arenicola (Photographer: R.R. Klopper)

Etymology

arenicola: For the preferred sandy habitat, from Latin 'arena' (sand), '-cola' (inhabiting).

Common names

bont-o-t'korrie [71] [Afrikaans]
strandveldaalwyn [9, 55] [Afrikaans]

Aloe argenticauda Merxm. & Giess

Etymology

argenticauda: For the peduncle covered with large, silvery bracts, from Latin 'argenteus' (silvery), 'cauda' (tail).

Common names

silver-tailed aloe [55] [English]

Aloe argyrostachys Lavranos, Rakouth & T.A.McCoy

Etymology

argyrostachys: For the silvery aspect of the flowers, from Greek 'argyro' (silver), 'stachys' (spike).

Aloe aristata Haw.

Synonyms

A. aristata Haw. var. *leiophylla* Baker
A. aristata Haw. var. *parvifolia* Baker
A. ellenbergeri Guillaumin
A. longiaristata Schult. & Schult.f.

Etymology

aristata: For the awn-like leaf tips, from Latin 'aristatus' (awned).

ellenbergeri: For Rene Ellenberger who first collected the taxon in Lesotho in 1920.

leiophylla: For the smooth leaves, from Greek 'leios' (smooth), 'phyllon' (leaf).

Aloe aristata (Photographer: N.R. Crouch)

longiaristata: For the awn-like leaf tips, from Latin 'longus' (long), 'aristatus' (awned).
parvifolia: For the smaller leaves, from Latin 'parvus' (small), '-folia' (leaved).

Common names

baard-aalwyn [106] [Afrikaans]
guinea-fowl aloe [55, 96] [English]
lace aloe [56] [English]
langnaaldaalwyn [55, 65] [Afrikaans]
lekhala-le-lenyenyane [58, 96] [Sotho, Southern]
lekhalana [58, 96] [Sotho, Southern]
lekhalana-la-balimo [58] [Sotho, Southern]
long-awned aloe [55, 65] [English]
ntehiseng [58] [Sotho, Southern]
serelei [53, 55, 123] [Sotho, Southern]
sereleli [58, 95] [Sotho, Southern]
tarentaalaalwyn [55, 95, 96] [Afrikaans]
torch-plant [56] [English]
umathithibala [35, 55, 95, 96] [Zulu]

Aloe armatissima Lavranos & Collen.

Etymology

armatissima: For the prominent marginal teeth of the leaves, from Latin 'armatus' (armed), superlative.

Aloe asperifolia A.Berger

Aloe asperifolia (Photographer: M. Koekemoer)

Etymology

asperifolia: For the rough leaves, from Latin 'asper' (rough), '-folius' (leaved).

Common names

||gores [33] [Nama]
aukoreb [55] [Kwangali] [122] [Nama]
heksekringe [34, 99] [Afrikaans]
kraalaalwyn [9, 33, 34, 55, 99] [Afrikaans]
rauhblättrige aloe [33, 34, 99] [German]

Aloe aufensis T.A.McCoy

Etymology

aufensis: For Mt Jebel Auf in Saudi Arabia, where it was discovered.

Aloe aurelienii J.-B.Castillon

Etymology

aurelienii: For Aurélien Castillon, grandson of the author.

Common names

vaho [29] [Malagasy]

Aloe austroarabica T.A.McCoy & Lavranos

Etymology

austroarabica: For the occurrence in southern Saudi Arabia, from Latin 'auster' (south), 'arabicus' (Arabian).

Aloe babatiensis Christian & I.Verd.

Aloe babatiensis (Photographer: SANBI, G.W. Reynolds)

Etymology

babatiensis: For the presumed occurrence at Babati, in Tanzania.

Common names

karangheri [55, 102] [Iraqw]

Aloe bakeri Scott-Elliot

Aloe bakeri (Photographer: SANBI, D.S. Hardy)

Synonyms

Guillauminia bakeri (Scott-Elliot) P.V.Heath

Etymology

bakeri: For John G. Baker (1834–1920), British botanist at the Royal Botanic Gardens, Kew.

Aloe ballii Reynolds var. *ballii*

Etymology

ballii: For John S. Ball (fl. 1954–1974), Zimbabwean forestry officer.

Aloe ballii var. ballii (Photographer: SANBI, D.C.H. Plowes)

Common names

John Ball's cliff aloe [7, 55, 126] [English]

Aloe ballii Reynolds var. *makurupiniensis* Ellert

Etymology

ballii: For John S. Ball (fl. 1954–1974), Zimbabwean forestry officer.

makurupiniensis: For the occurrence near Makurupini River, in Zimbabwe/Mozambique.

Aloe ballyi Reynolds

Etymology

ballyi: For Dr Peter R.O. Bally (1895–1980), Swiss botanist at the Coryndon Museum, Nairobi, and resident in Kenya, who travelled widely in East Africa.

Aloe ballyi (Photographer: SANBI, G.W. Reynolds)

Common names

kipapa [55] [Dawida]
rat aloe [6, 55, 102] [English]
rora [55] [Unspecified language]

Aloe barberae Dyer

Synonyms

A. bainesii Dyer
A. bainesii Dyer var. *barberae* (Dyer) Baker
A. zeyheri Baker (nom. illegit.)

Etymology

bainesii: For John Thomas Baines (1820–1875), English artist and explorer, active mainly in South Africa.

barberae: For Mary E. Barber (née Bowker), (1818–1899), English writer, painter and naturalist whose parents emigrated to South Africa in 1820. She was one of the pioneer plant collectors in South Africa and introduced the plant to British horticulture[116].

Aloe ballii var. *ballii*

Aloe barberae (Photographer: G. Nichols)

zeyheri: For Karl L.P. Zeyher (1799–1858), German naturalist and botanical explorer in South Africa.

Common names

boomaalwyn [53, 55, 62, 94, 99, 108, 109, 119, 123, 124] [Afrikaans]
boomalwyn [99] [Afrikaans]
impondondo [53, 55, 99] [Zulu]
indlabendlazi [55, 99] [Zulu]
inhlaba [64] [Swati]
inkalane unkulu [50, 53, 55, 99, 123] [Zulu]
inkalane-enkulu [94, 99, 119] [Zulu]
mangana grande [37] [Unspecified language]
mikaalwyn [53, 55, 62, 99, 108, 109, 123, 124] [Afrikaans]
tree aloe [9, 53, 55, 64, 94, 99, 109, 119, 123, 124] [English]
umgxwala [48, 55, 99] [Zulu]
umgzwala [53, 55] [Zulu]
umhlabandlanzi [94, 99, 119] [Zulu]
umhlalampofu [94, 99, 119] [Zulu]
umpondonde [94, 99, 119] [Zulu]
xiteti [53, 55] [Unspecified language]

Aloe barbertoniae Pole-Evans

Aloe barbertoniae (Photographer: SANBI, G.W. Reynolds)

Etymology

barbertoniae: For the occurrence at Barberton, in South Africa.

Common names

Barberton aalwyn [86] [Afrikaans]
Barberton aloe [86] [English]

Aloe bargalensis Lavranos

Aloe bargalensis (Photographer: SANBI, J.J. Lavranos)

Etymology

bargalensis: For the occurrence at Bargal, in Somalia.

Aloe belavenokensis (Rauh & Gerold) L.E.Newton & G.D.Rowley

Synonyms

Lomatophyllum belavenokense Rauh & Gerold

Etymology

belavenokense/*belavenokensis*: For the occurrence near Belavenoka, in Madagascar.

Aloe bella G.D.Rowley

Synonyms

A. pulchra Lavranos (nom. illegit.)

Etymology

bella: For its beauty, from Latin 'bellus' (beautiful).
pulchra: For its beauty, from Latin 'pulcher' (beautiful).

Aloe bellatula Reynolds

Aloe bellatula (Photographer: SANBI, P. Joffe)

Synonyms

Guillauminia bellatula (Reynolds) P.V.Heath

Etymology

bellatula: For its beauty, from Latin 'bellus' (beautiful), diminutive.

Aloe berevoana Lavranos

Etymology

berevoana: For the occurrence near Berevo, in Madagascar.

Aloe bernadettae J.-B.Castillon

Etymology

bernadettae: For Bernadette Castillon, horticulturalist at La Réunion, expert cultivator of Madagascan succulents.

Aloe bertemariae Sebsebe & Dioli

Etymology

bertemariae: For Berte Marie Ulvester, wife of Dr Maurizio Dioli, Italian veterinary officer in Ethiopia.

Aloe betsileensis H.Perrier

Etymology

betsileensis: For the occurrence at Betsileo, in Madagascar.

Aloe bicomitum L.C.Leach

Synonyms

A. venusta Reynolds (nom. illegit.)

Etymology

bicomitum: For Dr G.W. Reynolds (1895–1967) and Neil R. Smuts (1898–1963) who travelled together in search of

plants, from Latin 'bi-' (two), 'comitor' (accompany).
venusta: Refers to its beauty, from Latin.

Aloe boiteaui Guillaumin

Synonyms

Leemea boiteaui (Guillaumin) P.V.Heath

Etymology

boiteaui: For Pierre L. Boiteau (1911–1980), French botanist in Madagascar and curator of the Botanical Garden in Antananarivo.

Aloe boscawenii Christian

Etymology

boscawenii: For Lieut.-Col. Mildmay Thomas Boscawen (1892–1958), English military officer who became a sisal (*Agave sisalana*) grower in Tanzania after World War I, where he developed a fine garden of succulent plants.

Aloe bosseri J.-B.Castillon

Etymology

bosseri: For Jean-M. Bosser (1922–), French botanist and agronomical engineer, director of Office de la Recherche Scientifique et Technique d'Outre-Mer (ORSTOM) at Antananarivo, in Madagascar.

Aloe bowiea Schult. & Schult.f.

Synonyms

A. bourea Schult. & Schult.f.
Bowiea africana Haw.
Chamaealoe africana (Haw.) A.Berger

Etymology

africana: For the occurrence in Africa.
bourea: Unresolved application; possibly an orthographic variant of *bowiea*.

Aloe bowiea (Photographer: J. Kirkel)

bowiea: For James Bowie (1789–1869), English horticulturalist and botanical collector in South Africa, who first collected the species[115].

Common names

Coega aloe [55, 109] [English]
Coega-aalwyn [55, 109] [Afrikaans]
kleinaalwyn [55, 109] [Afrikaans]

Aloe boylei Baker

Aloe boylei (Photographer: G. Nichols)

Synonyms

A. boylei Baker subsp. *major* Hilliard & B.L.Burtt

Etymology

boylei: For Mr F. Boyle who collected the first material in Natal (now KwaZulu-Natal) in 1891, with Mr Allison, who sent it to England where the plant was described.
major: For the size, from Latin 'magnus' (great), comparative.

Common names

broad-leaved grass aloe [95, 96] [English]
incothobe [47, 48] [Zulu]
isiphukuthwane [95, 96] [Zulu]
isiphukutwane [55, 123] [Zulu]
isiphuthumane [95] [Zulu]
isiputumane [55, 123] [Zulu]
ocothobe [95, 96] [Zulu]

Aloe brachystachys Baker

Synonyms

A. lastii Baker
A. schliebenii Lavranos

Etymology

brachystachys: For the short inflorescence, from Greek 'brachys' (short), 'stachys' (spike).
lastii: For J.T. Last, who collected the plant in 1885 and sent it to the Royal Botanic Gardens, Kew where it flowered.
schliebenii: For Hans Joachim E. Schlieben (1902–1975), German botanist, who collected the type in Tanzania.

Aloe branddraaiensis Groenew.

Etymology

branddraaiensis: For the occurrence at Branddraai in Mpumalanga Province, South Africa.

Aloe branddraaiensis (Photographer: SANBI, P. Joffe)

Common names

Branddraai aloe [55, 109] [English]
Branddraai-aalwyn [55, 109] [Afrikaans]

Aloe brandhamii S.Carter

Etymology

brandhamii: For Dr Peter E. Brandham (1937–), British plant geneticist at the Jodrell Laboratory, Royal Botanic Gardens, Kew with a strong interest in *Aloe*.

Aloe brevifolia Mill. var. *brevifolia*

Synonyms

A. brevifolia Mill. var. *postgenita* (Schult. & Schult.f.) Baker
A. brevioribus Mill.
A. perfoliata L. var. δ L.
A. perfoliata L. var. ζ L.
A. postgenita Schult. & Schult.f.
A. prolifera Haw.
A. prolifera Haw. var. *major* Salm-Dyck

Aloe boylei

Aloe brevifolia var. brevifolia (Photographer: G.F. Smith)

Etymology

brevifolia: For the short leaves, from Latin 'brevis' (short), '-folius' (leaved).
brevioribus: For the short leaves, from Latin 'brevis' (short).
major: For the size, from Latin 'magnus' (great), comparative.
postgenita: Application obscure, from Latin 'post' (behind or after), 'genitus' (produced, born).
perfoliata: For the stem passing through the leaves, i.e. the leaves are amplexicaul, from Latin 'per' (through), 'folia' (leaf).
prolifera: For being proliferous, from Latin.

Common names

aanteel-aalwyn [55] [Afrikaans]
duine-aalwyn [55, 83, 108, 123] [Afrikaans]
kleinaalwyn [55, 62, 83, 108, 123] [Afrikaans]

Aloe brevifolia var. *brevifolia*

Aloe brevifolia Mill. var. *depressa* (Haw.) Baker

Synonyms

A. brevifolia Mill. var. *serra* (DC.) A.Berger
A. depressa Haw.
A. serra DC.

Etymology

brevifolia: For the short leaves, from Latin 'brevis' (short), '-folius' (leaved).
depressa: Because the rosette appears vertically flattened, but it has also been suggested that leaves are less thick (flattened adaxially) than those of the typical variety[53]. From Latin 'depressus' (depressed).
serra: For the serrate leaves, from Latin 'serra' (saw).

Aloe breviscapa Reynolds & P.R.O.Bally

Etymology

breviscapa: For the short inflorescence, from Latin 'brevis' (short), 'scapus' (scape).

Aloe broomii Schönland var. *broomii*

Aloe broomii var. broommii (Photographer: J. Kirkel)

Etymology

broomii: For Dr Robert Broom (1866–1951), Scottish physician and palaeoanthropologist who emigrated to South Africa in 1896 and was the first to collect the plant in 1905.

Common names

berg alwyn [56] [Afrikaans]
bergaalwee [9, 62, 98, 108] [Afrikaans]
bergaalwyn [53, 55, 98, 106, 108, 123] [Afrikaans]
mountain aloe [106] [English]
slangaalwyn [55, 62, 98, 108, 123] [Afrikaans]

Aloe broomii Schönland var. *tarkaensis* Reynolds

Aloe broomii var. *tarkaensis* (Photographer: N.R. Crouch)

Aloe broomii var. broomii

Etymology

broomii: For Dr Robert Broom (1866–1951), Scottish physician and palaeoanthropologist who emigrated to South Africa in 1896 and was the first to collect the plant in 1905.

tarkaensis: For the occurrence near Tarkastad in the Eastern Cape Province, South Africa.

Aloe brunneodentata Lavranos & Collen.

Etymology

brunneodentata: For the brown marginal teeth of the leaves, from Latin 'brunneo' (brown), 'dentatus' (toothed).

Aloe brunneostriata Lavranos & S.Carter

Etymology

brunneostriata: For the striate leaves, from Latin 'brunneo' (brown), 'striatus' (striate).

Aloe bruynsii P.I.Forst.

Etymology

bruynsii: For Dr Peter V. Bruyns (1957–), South African mathematician and succulent plant specialist.

Aloe buchananii Baker

Aloe buchananii (Photographer: SANBI, G.W. Reynolds)

Etymology

buchananii: For John Buchanan (1821–1903), Scottish clergyman, resident in South Africa in 1861–1877.

Common names

maluwa [102] [Nyanja]

Aloe buchlohii Rauh

Etymology

buchlohii: For Prof. Günther Buchloh (1923–), German botanist in Stuttgart, who collected plants with Prof. Werner Rauh in Madagascar in 1961.

Aloe buettneri A.Berger

Aloe buettneri (Photographer: SANBI, G.W. Reynolds)

Synonyms

A. barteri Baker, p.p.
A. congolensis De Wild. & T.Durand

Etymology

barteri: For Charles Barter (fl. 1857–1859) British gardener, foreman of the Regent's Park gardens of the Royal Botanic Society, London who joined the second Niger Expedition of W. Baikie and collected the type.
buettneri: For Prof. Oscar A.R. Büttner (1858–1927), German botanist, head of a research station in Togo in 1890–1891, later professor in Berlin, who collected the plant.
congolensis: For the occurrence in the Democratic Republic of the Congo.

Common names

balli nyibi [55] [Fula] [38] [Fulfulde, Adamawa] [27] [Fulfulde, Nigerian]
balli nyiwa [55] [Fula] [38] [Fulfulde, Adamawa] [27] [Fulfulde, Nigerian]
bamalagba [27] [Maninkakan, Western]
bangio fauru [27] [Pulaar]
baza [55] [Bamanankan] [55, 102] [Senoufo]
beskore [55] [Fula]
boi [55, 102] [Bamanankan]
gbadu [27, 38, 55] [Tiv]
hántsàr gíiwáá [27] [Hausa]
hantsar giwa [38] [Hausa]
kábàr gíiwáá [27] [Hausa]
kabar giwa [38, 55] [Hausa]
kabargiwa [55] [Hausa]
kadio [27] [Maninkakan, Western]
kandio [27] [Maninkakan, Western]
kibela [55] [Kimbundu]
kikalanga kibela [55, 102] [Kimbundu]
kikalangu [55] [Kimbundu]
kpipiko [27] [Kulango, Bouna]
magno gu dondialé [27] [Cerma]
maposo [55, 102] [Mwani]
men-tipa [27] [Mòoré]
moda [38] [Hausa]
móódáá [27] [Hausa]
na-pug-maande [27] [Mòoré]
nimbéléké [27] [Senoufo, Djimini]
nsesareso abrobe [27, 38, 55] [Abron]
omvi [27] [Gbagyi]
omwi [38, 55] [Gbari]

pinangru [27] [Senoufo, Tagwana]
ramadanhi [55] [Fula]
sereberebe [27, 38, 55] [Akan]
sere-berebe [1, 55] [Unspecified language]
sinzé toro [27] [Jula]
sogoba bu [27] [Mandinka]
sogoba hu [27] [Bambara]
sogoba ku [55] [Bamanankan]
tienkara sansugu [27] [Karaboro, Eastern]
tulédagbla [27] [Unspecified language]
West African aloe [27] [English]
wudie [27] [Baoulé]
zààbóó [27] [Hausa]
zaabuwaa [27] [Hausa]
zabo [38, 55] [Hausa]
zaboko [55] [Hausa]
zabon dafi [55] [Hausa]
zabuwa [38, 55] [Hausa]
zapua disum [55] [Kusaal]

Aloe buhrii Lavranos

Aloe buhrii (Photographer: SANBI, P. Joffe)

Etymology

buhrii: For Elias A. Buhr, a farmer near Nieuwoudtville in South Africa, who first collected the species.

Aloe bukobana Reynolds

Etymology

bukobana: For the occurrence near Bukoba, in Tanzania.

Common names

nkaka [55, 102] [Haya]

Aloe bukobana (Photographer: SANBI, G.W. Reynolds)

Aloe bulbicaulis Christian

Aloe bulbicaulis (Photographer: R.R. Klopper)

Synonyms

A. trothae A.Berger

Etymology

bulbicaulis: For the bulbous base of the plant, from Latin 'bulbus' (bulb), 'caulis' (stem).
trothae: For Lothar von Trotha (1848–1920), German soldier in German East Africa in 1894–1897, who collected the type in Tanzania.

Common names

chinthembwe [82] [Nyanja] [82] [Tumbuka]
khuzi [82] [Ngoni]
lichongwe [82] [Yao]

Aloe bulbillifera H.Perrier var. *bulbillifera*

Etymology

bulbillifera: For the small bulbils that develop on the inflorescence, from Latin 'bulbilla' (small bulb), '-fer' (bearing).

Aloe bulbillifera H.Perrier var. *paulianae* Reynolds

Etymology

bulbillifera: For the small bulbils that develop on the inflorescence, from Latin 'bulbilla' (small bulb), '-fer' (bearing).
paulianae: For L. Paulian (wife of R. Paulian, then Deputy Director of the Institut Scientifique de Madagascar) who first collected the plants.

Aloe bullockii Reynolds

Etymology

bullockii: For Arthur A. Bullock (1906–1980), British botanist at the Royal Botanic Gardens, Kew and specialist in Asclepiadaceae.

Aloe burgersfortensis Reynolds

Etymology

burgersfortensis: For the occurrence near Burgersfort in South Africa.

Common names

Burgersfort aloe [55, 109] [English]
Burgersfortaalwyn [55] [Afrikaans]
Burgersfort-bontaalwyn [55, 109] [Afrikaans]

Aloe bulbicaulis

Aloe burgersfortensis (Photographer: G.F. Smith)

Aloe bussei A.Berger

Aloe bussei (Photographer: C.S. Björa)

Synonyms

A. morogoroensis Christian

Etymology

bussei: For Dr W. Busse, German agricultural officer in Tanzania.
morogoroensis: For the occurrence at Morogoro, in Tanzania.

Aloe calcairophila Reynolds

Aloe calcairophila (Photographer: G. Orlando)

Synonyms

Guillauminia calcairophila (Reynolds) P.V.Heath

Etymology

calcairophila: For its ecological preference for lime, from the French 'calcaire' (lime) and Greek 'philos' (friend).

Aloe calidophila Reynolds

Etymology

calidophila: For the preference for hot sites, from Latin 'calidus' (hot) and Greek 'philos' (friend).

Aloe calidophila (Photographer: SANBI, G.W. Reynolds)

Common names

argeesaa sodu [40] [Oromo, West Central]
genenoo [40] [Hadiyya]
godole uta [40] [Wolaytta]
hargeisa sodu [55, 102] [Boran]
heejersaa [40] [Oromo, West Central]

Aloe cameronii Hemsl. var. *bondana* Reynolds

Aloe cameronii var. *bondana* (Photographer: M.J. Kimbeley)

Etymology

bondana: For the occurrence near Bonda Mission, in Zimbabwe.
cameronii: For Kenneth J. Cameron, Scottish planter in Malawi for the African Lakes Coorporation 1890–1903.

Common names

Bonda Ruwari aloe [126] [English]

Aloe cameronii Hemsl. var. *cameronii*

Etymology

cameronii: For Kenneth J. Cameron, Scottish planter in Malawi for the African Lakes Coorporation 1890–1903.

Common names

Cameron's Ruwari aloe [55, 126] [English]
chinthembwe [82] [Nyanja] [82] [Tumbuka]
khuzi [82] [Ngoni]
lichongwe [82] [Yao]

Aloe cameronii Hemsl. var. *dedzana* Reynolds

Etymology

cameronii: For Kenneth J. Cameron, Scottish planter in Malawi for the African Lakes Coorporation 1890–1903.
dedzana: For the occurrence on Dedza Mountain, in Malawi.

Common names

chinthembwe [82] [Nyanja] [82] [Tumbuka]
khuzi [82] [Ngoni]
lichongwe [82] [Yao]

Aloe camperi Schweinf.

Synonyms

A. abyssinica Salm-Dyck (nom. illegit.)
A. albopicta A.Berger

Aloe camperi (Photographer: H.M. Steyn)

A. eru A.Berger
A. eru A.Berger var. *cornuta* A.Berger
A. spicata Baker (nom. illegit.)

Etymology

abyssinica: For the occurrence in Abyssinia.
albopicta: For the white spots on the leaves, from Latin 'albus' (white), 'pictus' (painted).
camperi: For Manfredo Camperi, resident in Eritrea.
cornuta: For the marginal leaf teeth, from Latin 'cornutus' (horned).
eru: For the common name of the plant, 'eru', in Ethiopia.
spicata: For the long and densely-flowered spike-like inflorescences, from Latin 'spicatus' (spicate).

Common names

Camper's aloe [55, 109] [English]
erreh [55, 102] [Unspecified language]
genenoo [40] [Hadiyya]
groenaalwyn [55, 109] [Afrikaans]
iret [55, 102] [Unspecified language]
sanda 're [40] [Tigrigna]

Aloe canarina S.Carter

Etymology

canarina: For the canary-yellow colour of the flowers, from Latin.

Aloe candelabrum A.Berger

Aloe candelabrum (Photographer: G.F. Smith)

Etymology

candelabrum: For the appearance of the inflorescence, which resembles a candelabrum or candlestick, from Latin.

Common names

candelabra aloe [55, 94, 99, 108, 124] [English]
candelabrum aloe [55, 62] [English]
doringaalwyn [55, 62] [Afrikaans]
doringveldaalwyn [55, 108] [Afrikaans]
ikhala [99] [Xhosa]
imihlaba [99] [Zulu]
inhlaba [55, 94, 99] [Zulu]
inkalane [55, 94, 99] [Zulu]
kandelaaraalwyn [9, 94, 99, 124] [Afrikaans]
Natalse doringveldaalwyn [124] [Afrikaans]
umhlaba [49, 55, 99] [Zulu]

Aloe canis S.Lane

Aloe canis (Photographer: S.S. Lane)

Etymology

canis: For Theo Campbell-Barker, who discovered the taxon, from Latin 'canis' (dog), in allusion to dog's barking (Barker)[70].

Common names

chinthembwe [82] [Nyanja] [82] [Tumbuka]
khuzi [82] [Ngoni]
lichongwe [82] [Yao]

Aloe cannellii L.C.Leach

Etymology

cannellii: For Ian C. Cannell (fl. 1967–2003), Zimbabwean civil engineer who travelled and collected plants with L.C. Leach.

Aloe capitata Baker var. *angavoana* J.-P.Castillon

Etymology

angavoana: For the occurrence in the Angavo Mountains near Ankazobe, in Madagascar.
capitata: For the head-like inflorescence, from Latin 'capitatus' (capitate).

Aloe capitata Baker var. *capitata*

Aloe capitata var. *capitata* (Photographer: N.R. Crouch)

Synonyms

A. cernua Tod.

Etymology

capitata: For the head-like inflorescence, from Latin 'capitatus' (capitate).
cernua: For the drooping flowers, from Latin 'cernuus' (slightly drooping).

Common names

sahondra [55, 91] [Unspecified language]

Aloe capitata Baker var. *quartziticola* H.Perrier

Aloe capitata var. *quartziticola* (Photographer: R.R. Klopper)

Etymology

capitata: For the head-like inflorescence, from Latin 'capitatus' (capitate).
quartziticola: For the occurrence on quartzite rock, from English/French 'quartzite' and Latin '-cola' (inhabiting).

Aloe capitata Baker var. *silvicola* H.Perrier

Etymology

capitata: For the head-like inflorescence, from Latin 'capitatus' (capitate).
silvicola: For the occurrence in forests, from Latin 'sylva' (forest), '-cola' (inhabiting).

Aloe capmanambatoensis Rauh & Gerold

Etymology

capmanambatoensis: For the occurrence at Cap Manambato, in Madagascar.

Aloe carnea S.Carter

Etymology

carnea: For the flesh colour of the flower, from Latin 'carneus' (meat).

Aloe carolineae L.E.Newton

Etymology

carolineae: For Caroline Wheeler (1960–2000), wife of Charlie Wheeler, Kenya, both active in the conservation of Kenya's environment.

Aloe castanea Schönland

Aloe castanea (Photographer: J.E. Burrows)

Etymology

castanea: For the chestnut brown colour of the nectar[123] or of the flowers[53], from Latin 'castanea' (chestnut).

Common names

barolo [53, 55, 99] [Sotho, Northern]
cat's-tail aloe [55, 109, 123, 124] [English]
chestnut brown aloe [99] [English]
katstertaalwyn [55, 62, 108, 109, 123, 124] [Afrikaans]
sawupo [55] [Unspecified language]
suwopa [53, 99] [Sotho, Northern]

Aloe castellorum J.R.I.Wood

Synonyms

A. hijazensis Lavranos & Collen.

Etymology

castellorum: For the occurrence on historic fortress mountains, from Latin 'castellum' (castle).
hijazensis: For the occurrence in Hijaz Province, in Saudi Arabia.

Aloe castilloniae J.-B.Castillon

Etymology

castilloniae: For Bernadette Castillon, horticulturalist at La Réunion, expert cultivator of Madagascan succulents.

Aloe cataractarum T.A.McCoy & Lavranos

Etymology

cataractarum: For the occurrence near waterfalls, from Latin 'cataracta' (waterfall).

Aloe catengiana Reynolds

Etymology

catengiana: For the occurrence near Catengue, in Angola.

Common names

okandolle [55, 102] [Umbundu]

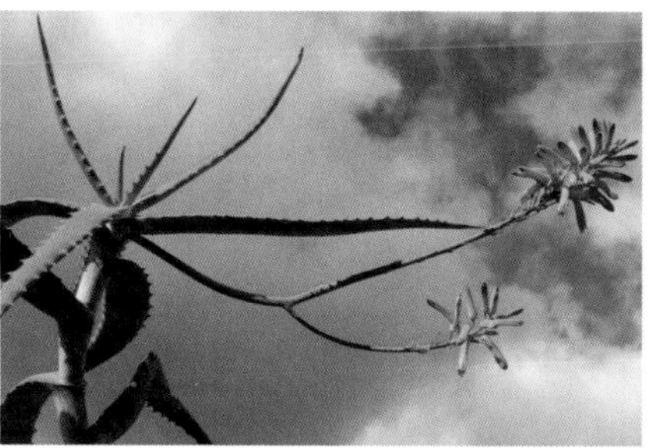
Aloe catengiana (Photographer: E.J. van Jaarsveld)

Aloe chabaudii Schönland var. *chabaudii*

Aloe chabaudii var. *chabaudii* (Photographer: M.J. Kimberley)

Synonyms

A. chabaudii Schönland var. *verekeri* Christian

Etymology

chabaudii: For John A. Chabaud, plant grower in Port Elizabeth, South Africa, in whose garden the original specimens flowered.

verekeri: For Mr L.S.A. Vereker, keen collector of Zimbabwean succulents, who first collected the species in 1931.

Common names

Chabaud's aloe [55, 95] [English]
chigiakia [8, 129] [Tswa]
chikowa [8, 103, 129] [Shona]
chinthembwe [82] [Nyanja] [82] [Tumbuka]
chinungu [8, 129] [Shona]
chinyangami [8, 103, 129] [Tonga]
chisongwe [55, 68, 69] [Unspecified language]
Chizarira escarpment aloe [126] [English]
cilombo [55] [Lozi]
cisongwe [55] [Nyanja]
dwala aloe [55, 126] [English]
gavakava [8, 45, 55, 129] [Shona]
godzongo [8, 129] [Shona]
grey aloe [55, 109] [English]
grysaalwyn [55, 109] [Afrikaans]
gweravana [8, 129] [Shona]
Hunyani range aloe [126] [English]
icena [8, 45, 129] [Ndebele]
ihlaba [55] [Ndebele]
inhlaba [95] [Swati] [95] [Zulu]
inkalane [95] [Zulu]
khuzi [82] [Ngoni]
lichongwe [82] [Yao]
khonje wa fisi [55, 68, 69] [Unspecified language]
khonje wa mbuzi [55, 68, 69] [Unspecified language]
llai [37] [Unspecified language]
madaka [68, 69] [Unspecified language] [55] [Nyanja]
mangana [37] [Unspecified language]
mangani [8] [Tswa]
nemba [69] [Unspecified language]
ngosiya [55, 68, 69] [Unspecified language]
nsenjere [55] [Zulu]
pozo [55] [Nyanja]
rumhangamhuno [8] [Shona]
ruvati [8] [Shona]
tshikhopha [74] [Venda]
Vereker's dwala aloe [55, 126] [English]
Zambezi gorges aloe [126] [English]

Aloe chabaudii Schönland var. *mlanjeana* Christian

Aloe chabaudii var. *mlanjeana* (Photographer: S.S. Lane)

Etymology

chabaudii: For John A. Chabaud, plant grower in Port Elizabeth, South Africa, in whose garden the original specimens flowered.
mlanjeana: For the occurrence on Mt Mlanje, in Malawi.

Common names

chinthembwe [82] [Nyanja] [82] [Tumbuka]
khuzi [82] [Ngoni]
lichongwe [82] [Yao]

Aloe challisii van Jaarsv. & A.E.van Wyk

Aloe challisii (Photographer: E.J. van Jaarsveld)

Etymology

challisii: For Mr Chris Challis, an *Aloe* and succulent enthusiast, who first collected the species while exploring a hiking trail in Verlorenkloof, in South Africa.

Aloe charlotteae J.-B.Castillon

Etymology

charlotteae: For Charlotte Castillon, granddaughter of the author.

Aloe cheranganiensis S.Carter & Brandham

Etymology

cheranganiensis: For the occurrence in the Cherangani Hills, in Kenya.

Aloe chlorantha Lavranos

Aloe chlorantha (Photographer: SANBI)

Etymology

chlorantha: For the green flower, from Greek 'chloros' (green), 'anthos' (flower).

Aloe chortolirioides A.Berger var. *chortolirioides*

Aloe chortolirioides var. *chortolirioides* (Photographer: SANBI, J. Onderstall)

Synonyms

A. boastii Letty
A. chortolirioides A.Berger var. *boastii* (Letty) Reynolds

Etymology

boastii: For Mr H.W. Boast, Deputy Assistant Commissioner, Pigg's Peak, in Swaziland, who collected it.
chortolirioides: For resembling the genus *Chortolirion*, particularly *C. angolense*, from Greek '-oides' (resembling).

Common names

grasaalwyn [9, 55] [Afrikaans]

Aloe chortolirioides A.Berger var. *woolliana* (Pole-Evans) Glen & D.S.Hardy

Aloe chortolirioides var. *woolliana* (Photographer: SANBI, G. Condy)

Synonyms

A. woolliana Pole-Evans

Etymology

chortolirioides: For resembling the genus *Chortolirion*, particularly *C. angolense*, from Greek '-oides' (resembling).
woolliana: For Mr Woolley, who lived in Barberton in the 1930s and collected the first specimen.

Common names

grasaalwyn [9] [Afrikaans]

Aloe christianii Reynolds

Aloe christianii (Photographer: SANBI, L.C. Leach)

Etymology

christianii: For H. Basil Christian (1871–1950), South African agriculturalist and amateur botanist who emigrated to Zimbabwe in 1911 and established a large private garden in 1914 that became the Ewanrigg National Park.

Common names

Basil Christian's aloe [55, 126] [English]
chinthembwe [82] [Nyanja] [82] [Tumbuka]
chizimamuliro [82] [Ngoni]
chizime [69] [Unspecified language]
iwani [69] [Unspecified language]
kamingaminga [55, 68] [Unspecified language]
khondje [37] [Unspecified language]
khuzi [69] [Unspecified language]
lichingwe [55] [Nyanja]
licongwe [55] [Nyanja]
nsenjere [55] [Nyanja]
pikwe [55, 68] [Unspecified language]
wantchile [37] [Unspecified language]

Aloe chrysostachys Lavranos & L.E.Newton

Synonyms

A. meruana Lavranos

Etymology

chrysostachys: For the yellow inflorescence, from Greek 'chrysos' (gold), 'stachys' (spike).
meruana: For the occurrence in Meru Game Reserve, in Kenya.

Aloe ciliaris Haw. var. *ciliaris*

Aloe ciliaris var. *ciliaris* (Photographer: E.J. van Jaarsveld)

Synonyms

A. ciliaris Haw. var. *flanaganii* Schönland

Etymology

ciliaris: For the fringe of cilia on the stem-clasping/amplexicaul leaf bases, from Latin 'ciliaris' (ciliate).
flanaganii: For Henri G. Flanagan (1861–1919), South African farmer interested in botany.

Common names

climbing aloe [56, 99] [English]
fringing broader-leaved aloe [55] [English]
ikalene [99] [Xhosa]
intelezi [99] [Xhosa]
sábila china [88] [Spanish]

Aloe ciliaris Haw. var. *redacta* S.Carter

Etymology

ciliaris: For the fringe of cilia on the stem-clasping/amplexicaul leaf bases, from Latin 'ciliaris' (ciliate).
redacta: For the limited range, or for the reduced pedicels and bracts, from Latin 'redactus' (reduced).

Aloe ciliaris Haw. var. *tidmarshii* Schönland

Synonyms

A. tidmarshii (Schönland) F.S.Mull. ex R.A.Dyer

Etymology

ciliaris: For the fringe of cilia on the stem-clasping/amplexicaul leaf bases, from Latin 'ciliaris' (ciliate).
tidmarshii: For Edwin Tidmarsh (1831–1915), curator of the Grahamstown Botanic Garden, who, in 1900, gave the plant to Dr Selmar Schönland, who described it.

Aloe ciliaris var. *ciliaris*

Aloe ciliaris var. *tidmarshii* (Photographer: G.F. Smith)

Common names

fiery climber [9, 55] [English]

Aloe cipolinicola (H.Perrier) J.-B.Castillon & J.-P.Castillon

Aloe cipolinicola (Photographer: S.E. Rakotoarisoa)

Synonyms

A. capitata Baker var. *cipolinicola* H.Perrier

Etymology

capitata: For the head-like inflorescence, from Latin 'capitatus' (capitate).
cipolinicola: For the occurrence on Cipolin limestones in Madagascar, from Latin '-cola' (inhabiting).

Common names

vahonona [100] [Malagasy]

Aloe citrea (Guillaumin) L.E.Newton & G.D.Rowley

Synonyms

Lompatophyllum citreum Guillaumin

Etymology

citrea/citreum: For the lemon yellow flower colour, from Latin 'citreus' (lemon like).

Aloe citrina S.Carter & Brandham

Etymology

citrina: For the yellow flower colour, from Latin 'citrinus' (lemon-yellow).

Aloe clarkei L.E.Newton

Etymology

clarkei: For Paul Clarke, English management consultant resident in Kenya in 1985–2001 who discovered the plants on a remote mountain.

Aloe classenii Reynolds

Etymology

classenii: For George A. Classen (1915–1982), Russian-born geologist resident in Kenya from 1948, who collected plants while traveling professionally as a hydrologist.

Aloe claviflora Burch.

Aloe claviflora (Photographer: J.C. Kruger)

Synonyms

A. decora Schönland
A. schlechteri Schönland

Etymology

claviflora: For the club-shaped flowers, from Latin 'clava' (club), '-florus' (flowered).
decora: For the appearance, from Latin 'decorus' (graceful).
schlechteri: For Max Schlechter (1874–1960), German trader and plant collector in South Africa, who collected the type.

Common names

aanteelaalwyn [53, 55, 59, 60, 108] [Afrikaans]
cannon aloe [123] [English]
jakkalsstert [55] [Afrikaans]
kanonaalwyn [55, 59, 60, 62, 98, 106, 108, 123] [Afrikaans]
katstert [60] [Afrikaans]
katstertaalwyn [55, 59] [Afrikaans]
kraalaalwyn [9, 53, 55, 59, 60, 62, 98, 106, 108, 123] [Afrikaans]
kraalaloe [55] [English]
laeraalwyn [55, 62, 108] [Afrikaans]

Aloe collenetteae Lavranos

Etymology

collenetteae: For Iris Sheila Collenette (1927–), English amateur botanist, well-known collector and researcher of Arabian succulents.

Aloe collina S.Carter

Aloe collina (Photographer: M.J. Kimberley)

Etymology

collina: For the preferred hilly habitat, from Latin 'collinus' (hill).

Aloe commixta A.Berger

Synonyms

A. gracilis Baker (nom. illegit.)
A. perfoliata L. var. *α* L.

Aloe commixta (Photographer: E.J. van Jaarsveld)

Etymology

commixta: From Latin 'commixtus' (mixed up) perhaps because it was previously known under an illegitimate name[42] or because of its occurrence in dense thickets of intermingled stems[123], or because it was first confused with *Aloe gracilis* or considered a variant of *A. striatula*[53].

gracilis: For the slender stems, from Latin 'gracilis' (slender).

perfoliata: For the stem passing through the leaves, i.e. the leaves are amplexicaul, from Latin 'per' (through), 'folia' (leaf).

Aloe comosa Marloth & A.Berger

Etymology

comosa: For being brush-like, referring to the leaves[123], or to the inflorescences[53], from Latin 'coma' (hair, mane, tuft).

Common names

aloe with hair-like tufts [99] [English]
Clanwilliam aloe [55, 123, 124] [English]
Clanwilliamaalwyn [9, 55, 124, 123] [Afrikaans]

Aloe comosa (Photographer: SANBI, G.W. Reynolds)

Aloe compressa H.Perrier var. *compressa*

Etymology

compressa: For the distichous (laterally compressed) leaf arrangements, from Latin 'compressus' (compressed).

Aloe compressa H.Perrier var. *paucituberculata* Lavranos

Etymology

compressa: For the distichous (laterally compressed) leaf arrangements, from Latin 'compressus' (compressed).

paucituberculata: For the sparsely tuberculate leaves, from Latin 'pauci' (few), 'tuberculatus' (tuberculate).

Aloe compressa H.Perrier var. *schistophila* H.Perrier

Etymology

compressa: For the distichous (laterally compressed) leaf arrangements, from Latin 'compressus' (compressed).

schistophila: For the preferred habitat on schistose rocks, from Greek 'schistos' (schist rock), 'phylos' (friend).

Aloe comptonii Reynolds

Aloe comptonii (Photographer: G.F. Smith)

Synonyms

A. mitriformis Mill. subsp. *comptonii* (Reynolds) Zonn.

Etymology

comptonii: For Prof. Robert H. Compton (1886–1979), British botanist in South Africa, Professor at the University of Cape Town and second director of the

National Botanic Gardens of South Africa at Kirstenbosch.

mitriformis: For the appearance of the rosette apex, i.e. shaped like a Bishop's cap, from Latin 'mitris' (mitre), '-formis' (shaped).

Common names

Compton's aloe [55, 109] [English]
Kleinkaroo-aalwyn [55, 109] [Afrikaans]

Aloe confusa Engl.

Etymology

confusa: For being confused, as the taxon was previously unknown, from Latin.

Aloe congdonii S.Carter

Etymology

congdonii: For Colin Congdon, British manager of a tea estate in Tanzania and amateur naturalist.

Aloe conifera H.Perrier

Aloe conifera (Photographer: SANBI, G.W. Reynolds)

Etymology

conifera: For the cone-like appearance of young inflorescences, from Latin 'conus' (cone), '-fer' (carrying).

Aloe cooperi Baker subsp. *cooperi*

Aloe cooperi subsp. *cooperi* (Photographer: G.F. Smith)

Synonyms

A. schmidtiana Regel

Etymology

cooperi: For Thomas Cooper (1815–1913), English plant collector working for W.W. Saunders, who collected plants in South Africa in 1859–1862, and rediscovered the species in 1860 (it was first discovered by William J. Burchell).

schmidtiana: For the director of the German nursery of Haage and Schmidt.

Aloe comptonii

Common names

Cooper's aloe [9, 55] [English]
inkalane [35, 55] [Zulu]
isiphukuthwane [55, 95] [Zulu]
isiphukutwane [53, 55] [Zulu]
isiphuthumana [52] [Zulu]
isiphuthumane [55, 95] [Zulu]
isipukutwane [55, 101] [Zulu]
isiputumane [53, 55, 101] [Zulu]
lisheshelu [95] [Swati]
nkalane [55] [Zulu]

Aloe cooperi Baker subsp. *pulchra* Glen & D.S.Hardy

Etymology

cooperi: For Thomas Cooper (1815–1913), English plant collector working for W.W. Saunders, who collected plants in South Africa in 1859–1862, and rediscovered the species in 1860 (it was first discovered by William J. Burchell).
pulchra: For its beauty, from Latin 'pulcher' (beautiful).

Common names

aloé-capim [4] [Portuguese]
grass aloe [4] [English]
jejeje [4] [Changa]

Aloe corallina I.Verd.

Etymology

corallina: For the flower colour, from Latin 'corallinus' (coral red, coral-like).

Aloe ×corderoyi A.Berger (*A. plicatilis* × *A. variegata*)

Etymology

corderoyi: For Justus Corderoy (1832–1911), English succulent plant grower.

Aloe cooperi subsp. *cooperi*

Aloe ×corifolia Pillans

Etymology

corifolia: For the leathery leaf texture, from Latin 'corium' (skin), '-folia' (leaved).

Aloe craibii Gideon F.Sm.

Aloe craibii (Photographer: E. van Wyk)

Etymology

craibii: For Charles Craib (fl. 1997–2003), enthusiatic amateur botanist from Johannesburg, South Africa[110].

Aloe crassipes Baker

Aloe crassipes (Photographer: SANBI, G.W. Reynolds)

Etymology

crassipes: Application obscure, from Latin 'crassus' (thick), 'pes' (foot).

Aloe cremnophila Reynolds & P.R.O.Bally

Etymology

cremnophila: For the habitat, from Greek 'kremnos' (cliff, slope), 'philos' (friend).

Aloe cryptoflora Reynolds

Etymology

cryptoflora: For the flowers which are hidden by the bracts, from Greek 'kryptos' (hidden, covered) and Latin '-florus' (flowered).

Aloe cryptopoda Baker

Aloe cryptopoda (Photographer: M.J. Kimberley)

Etymology

cryptopoda: Because the flower bases are covered by the large bracts that hide the flower pedicels, from Greek 'kryptos' (hidden, covered), 'podos' (foot).

Common names

chikowa [103] [Shona]
chinyangami [103] [Tonga]
chitembwe [55] [Nyanja]
chitupa [55, 68] [Unspecified language]
citembwe [55] [Nyanja]
citupa [55] [Nyanja]
Dr Kirk's aloe [55, 126] [English]
gave wamtchanga [82] [Sena]
geelaalwee [108] [Afrikaans]
geelaalwyn [55, 62, 108, 109, 123] [Afrikaans]
lai [37] [Unspecified language]
lichongwe [55] [Nyanja]
licongwe [55] [Nyanja]
mdyang'oma [82] [Ngoni]
ngafane [53, 55, 123] [Unspecified language]
spire aloe [55] [English]
yellow aloe [55, 109] [English]

Aloe cyrtophylla Lavranos

Etymology

cyrtophylla: For the distal halves of the leaves that are rolled back, from Greek 'kyrtos' (curved), 'phyllon' (leaf).

Aloe dabenorisana van Jaarsv.

Aloe dabenorisana (Photographer: E.J. van Jaarsveld)

Etymology

dabenorisana: For the occurrence in the Dabenoris Mountain range, in South Africa.

Aloe darainensis J.-P.Castillon

Etymology

darainensis: For the occurrence near Daraina, in Madagascar.

Aloe davyana Schönland

Aloe davyana (Photographer: J. Kirkel)

Synonyms

A. comosibracteata Reynolds
A. davyana Schönland var. *subolifera* Groenew.
A. greatheadii Schönland var. *davyana* (Schönland) Glen & D.S.Hardy
A. labiaflava Groenew.

Etymology

comosibracteata: For the fleshy bracts in an imbricate tuft at the top of the raceme, from Latin 'coma' (hair, mane, tuft), 'bracteatus' (with bracts).
davyana: For Dr Joseph Burtt Davy (1870–1940), British botanist working in South Africa in 1903–1919, chief of the Division of Botany, Department of Agriculture.
greatheadii: For Dr J.B. Greathead who collected the type with Dr S. Schönland.
labiaflava: For the yellow perianth segments, from Latin 'labium' (lip), 'flavus' (yellow).
subolifera: For forming suckers, from Latin 'soboles' (branches).

Common names

bontaalwyn [55, 109] [Afrikaans]
bontblaar aalwyn [9, 55] [Afrikaans]
bontblaaraalwyn [125] [Afrikaans]
kgopane [53, 55, 123] [Tswana]
kleinaalwyn [55, 62, 108, 109, 123, 125] [Afrikaans]
kxophane [55, 125] [Tswana]
oorbeweidingsaalwyn [55] [Afrikaans]
spotted aloe [55, 109] [English]
Transvaal aalwyn [55] [Afrikaans]
Transvaalaalwyn [55, 123] [Afrikaans]

Aloe dawei A.Berger

Aloe dawei (Photographer: SANBI, G.W. Reynolds)

Aloe dabenorisana

Synonyms

A. beniensis De Wild.
A. pole-evansii Christian

Etymology

beniensis: For the occurrence near Beni, in the Democratic Republic of Congo.
dawei: For Morley T. Dawe (1880–1943), British forester in Uganda and curator of the Entebbe Botanical Garden.
pole-evansii: For Dr Illtyd B. Pole-Evans (1877–1968), botanist in South Africa.

Common names

kakarutanga [55, 102] [Tooro]
kokorutanga [55, 102] [Nyoro]
likokho [55] [Swahili]
lilinakha [55] [Swahili]
tangaratwe [55, 102] [Nandi]
tangaratwet [6] [Kipsigis] [55] [Nandi]

Aloe debrana Christian

Synonyms

A. berhana Reynolds

Etymology

berhana: For the occurrence at Debre Berhan, in Ethiopia.
debrana: For the occurrence at Debre Berhan, in Ethiopia.

Common names

argeesaa [40] [Oromo, West Central]
heejersaa [40] [Oromo, West Central]
merarie [40] [Amharic]

Aloe decaryi Guillaumin

Etymology

decaryi: For Raymond Decary (1891–1973), French financial administrator, botanist and plant collector in Madagascar in 1916–1944.

Aloe decorsei H.Perrier

Etymology

decorsei: For Dr Gaston-J. Decorse (1873–1907), French botanist and entomologist, collector in Madagascar in 1898–1900.

Aloe decumbens (Reynolds) van Jaarsv.

Synonyms

A. gracilis Haw. var. *decumbens* Reynolds

Etymology

decumbens: For the decumbent habit, from Latin.
gracilis: For the slender stems, from Latin 'gracilis' (slender).

Common names

inhlaba yentaba [55] [Zulu]
rankaalwyn [55] [Afrikaans]

Aloe decurva Reynolds

Etymology

decurva: For the orientation of the inflorescences, from Latin 'decurvus' (decurved).

Aloe deinacantha T.A.McCoy, Rakouth & Lavranos

Etymology

deinacantha: For the strong spines on the leaf margins, from Greek 'deinos' (dreadful, terrible), 'akantha' (thorn, spine).

Aloe ×*deleuili* Hort.

(*A. elegans* × *A. ferox*)

Etymology

deleuili: Unresolved application.

Aloe delphinensis Rauh

Etymology

delphinensis: For the occurrence near Fort Dauphin, in Madagascar, from Latin, of the dolphin, French 'dauphin'.

Aloe deltoideodonta Baker subsp. *amboahangyensis* Rebmann

Etymology

amboahangyensis: For the occurrence near Amboahangy, in Madagascar.
deltoideodonta: For the leaf marginal teeth, from Greek 'deltoides' (delta-shaped), 'odous, odontos' (tooth).

Aloe deltoideodonta Baker var. *brevifolia* H.Perrier

Etymology

brevifolia: For the short leaves, from Latin 'brevis' (short), '-folius' (leaved).
deltoideodonta: For the leaf marginal teeth, from Greek 'deltoides' (delta-shaped), 'odous, odontos' (tooth).

Aloe deltoideodonta Baker var. *candicans* H.Perrier forma *candicans*

Etymology

candicans: For the bract colour, from Latin 'candicans' (becoming white).
deltoideodonta: For the leaf marginal teeth, from Greek 'deltoides' (delta-shaped), 'odous, odontos' (tooth).

Aloe deltoideodonta Baker var. *candicans* H.Perrier forma *latifolia* H.Perrier

Etymology

candicans: For the bract colour, from Latin 'candicans' (becoming white).
deltoideodonta: For the leaf marginal teeth, from Greek 'deltoides' (delta-shaped), 'odous, odontos' (tooth).
latifolia: For the wide leaves, from Latin 'latus' (broad), '-folius' (leaved).

Aloe deltoideodonta Baker var. *candicans* H.Perrier forma *longifolia* H.Perrier

Etymology

candicans: For the bract colour, from Latin 'candicans' (becoming white).
deltoideodonta: For the leaf marginal teeth, from Greek 'deltoides' (delta-shaped), 'odous, odontos' (tooth).
longifolia: For the length of the leaves, from Latin 'longus' (long), '-folius' (leaved).

Aloe deltoideodonta Baker var. *deltoideodonta*

Synonyms

A. rossii Tod.

Etymology

deltoideodonta: For the leaf marginal teeth, from Greek 'deltoides' (delta-shaped), 'odous, odontos' (tooth).
rossii: For Ermanno Ross, the author's assistant at the Palermo botanical garden.

Common names

vahombato [55, 90] [Unspecified language]
vahongarana [55, 90] [Unspecified language]

Aloe deltoideodonta Baker var. *fallax* J.-B.Castillon

Etymology

deltoideodonta: For the leaf marginal teeth, from Greek 'deltoides' (delta-shaped), 'odous, odontos' (tooth).

fallax: Because it was thought to be another species, from Latin 'fallax' (deceptive).

Aloe deltoideodonta Baker var. *intermedia* H.Perrier

Etymology

deltoideodonta: For the leaf marginal teeth, from Greek 'deltoides' (delta-shaped), 'odous, odontos' (tooth).
intermedia: For the relationship to other taxa, from Latin 'intermedius' (intermediate).

Aloe deltoideodonta Baker var. *ruffingiana* (Rauh & Petignat) J.-B.Castillon & J.-P.Castillon

Synonyms

A. deltoideodonta Baker subsp. *esomonyensis* Rebmann
A. ruffingiana Rauh & Petignat

Etymology

deltoideodonta: For the leaf marginal teeth, from Greek 'deltoides' (delta-shaped), 'odous, odontos' (tooth).
esomonyensis: For the occurrence near Esomony, in Madagascar.
ruffingiana: For Dr E. Ruffing, German physician working in Madagascar.

Aloe descoingsii Reynolds subsp. *augustina* Lavranos

Etymology

augustina: For the occurrence near St Augustin, in Madagascar.
descoingsii: For Dr Bernard M. Descoings (1931–), French botanist and specialist on Madagascan plant diversity.

Aloe descoingsii Reynolds subsp. *descoingsii*

Aloe descoingsii subsp. *descoingsii* (Photographer: SANBI, W. Rauh)

Synonyms

Guillauminia descoingsii (Reynolds) P.V.Heath

Etymology

descoingsii: For Dr Bernard M. Descoings (1931–), French botanist and specialist on Madagascan plant diversity.

Aloe desertii A.Berger

Etymology

desertii: For the occurrence in the desert, from Latin.

Aloe ×desmetiana Baker
(*A. humilis* × *A. variegata*)

Etymology

desmetiana: Application obscure, possibly for Louis De Smet (1813–1887), Belgian horticulturalist and nurseryman.

Aloe dewetii Reynolds

Aloe dewetii (Photographer: SANBI, G.W. Reynolds)

Etymology

dewetii: For J.F. de Wet, Headmaster of Vryheid Junior School, South Africa, who brought the species to the attention of the author.

Common names

De Wet's aloe [95] [English]
inhlaba [95] [Swati]

Aloe dewinteri Giess

Etymology

dewinteri: For Dr Bernard de Winter (1924–), botanist and director of the then Botanical Research Institute (now SANBI) in Pretoria, South Africa, who collected one of the first plants from which the description was made.

Common names

Sesfonteinaalwyn [55, 59, 60] [Afrikaans]

Aloe dhufarensis Lavranos

Etymology

dhufarensis: For the occurrence in the Dhufar Province, in Oman.

Aloe dichotoma Masson

Aloe dichotoma (Photographer: J.C. Kruger)

Synonyms

A. dichotoma Masson var. *montana* (Schinz) A.Berger
A. montana Schinz

A. ramosa Haw.
Rhipidodendron dichotomum (Masson) Willd.

Etymology

dichotoma/dichotomum: For the branching of the stems, from Latin 'dichotomus' (dichotomous, division in pairs, forked).
montana: For the habitat in mountains, from Latin 'montanus' (mountain).
ramosa: For the branching of the stems, from Latin 'ramosus' (branched).

Common names

||garas [33, 36, 60, 75] [Nama]
boskokerboom [55] [Afrikaans]
choje [55, 108] [Unspecified language]
die dikke [53, 55] [Afrikaans]
garab [55] [Nama]
garas [53, 55, 99] [Nama]
köcherbaum [33, 36, 60, 75] [German]
kokerbaum [75] [German]
kokerboom [9, 25, 33, 36, 53, 55, 59, 60, 62, 75, 98, 99, 108, 109, 123, 124] [Afrikaans]
quiver tree [9, 25, 33, 36, 53, 55, 59, 60, 62, 75, 99, 108, 109, 123, 124] [English]

Aloe dichotoma

Aloe dinteri A.Berger

Aloe dinteri (Photographer: P.J.D. Winter)

Etymology

dinteri: For Prof. Moritz Kurt Dinter (1868–1945), German botanist famous for his explorations in Namibia, who discovered the plants in 1912.

Aloe diolii L.E.Newton

Etymology

diolii: For Dr Maurizio Dioli, Italian veterinary officer resident in Kenya, later in Ethiopia.

Aloe distans Haw.

Synonyms

A. brevifolia (Aiton) Haw. (nom. illegit.)
A. mitriformis Mill. subsp. *distans* (Haw.) Zonn.
A. mitriformis Mill. var. *angustior* Lam.
A. mitriformis Mill. var. *brevifolia* (Aiton) W.T.Aiton
A. mitriformis Mill. var. *humilior* Willd.
A. perfoliata L. var. *brevifolia* Aiton
A. reflexa van Marum ex Steud.

Etymology

angustior: For the narrower leaves, from Latin 'angustus' (narrow), comparative.
brevifolia: For the short leaves, from Latin 'brevis' (short), '-folius' (leaved).
distans: For the long internodes (distant), from Latin.
humilior: For being smaller, from Latin 'humilis' (low, modest), comparative.
mitriformis: For the appearance of the rosette apex, i.e. shaped like a Bishop's cap, from Latin 'mitris' (mitre), '-formis' (shaped).
perfoliata: For the stem passing through the leaves, i.e. the leaves are amplexicaul, from Latin 'per' (through), 'folia' (leaf).
reflexa: Probably for the leaves, from Latin 'reflexus' (reflexed).

Common names

short-leaved aloe [55] [English]
strandaalwyn [9, 55] [Afrikaans]

Aloe divaricata A.Berger var. *divaricata*

Aloe divaricata var. *divaricata* (Photographer: J.-P. Castillon)

Synonyms

A. vaotsohy Decorse & Poiss.

Etymology

divaricata: For the branching of the inflorescence, spreading, divaricate, from Latin.
vaotsohy: For the common name 'vahotsohy'.

Common names

vahomafaitra [55] [Tandroy-Mahafaly Malagasy]
vahona [55] [Tandroy-Mahafaly Malagasy]
vahonmafaitra [91] [Unspecified language]
vahonomafaitra [55, 90] [Unspecified language]
vahontsohy [55, 91, 90] [Unspecified language]
vahotsohy [90, 91] [Unspecified language] [55] [Tandroy-Mahafaly Malagasy]
vohandranjo [55] [Unspecified language]

Aloe divaricata A.Berger var. *rosea* (Decary) Reynolds

Synonyms

A. vaotsohy Decorse & Poiss. var. *rosea* Decary

Etymology

divaricata: For the branching of the inflorescence, spreading, divaricate, from Latin.
rosea: For the rose-pink flowers, from Latin 'roseus' (rose-like).
vaotsohy: For the common name 'vahotsohy'.

Aloe djiboutiensis T.A.McCoy

Etymology

djiboutiensis: For the occurrence in Djibouti.

Aloe doddsiorum T.A.McCoy & Lavranos

Etymology

doddsiorum: For Anthony and Maria Dodds who conducted fieldwork, and made many discoveries, in Kenya.

Aloe dominella Reynolds

Aloe dominella (Photographer: SANBI)

Etymology

dominella: Application obscure, may refer to the species being locally dominant in small areas[123], from Latin 'dominus' (lord, master), or it could be a corruption of 'dominilla' (the lady of the house) because the type was collected on a farm owned by a Miss Quested[53], from Latin 'domina' (lady, mistress), diminutive.

Aloe dorotheae A.Berger

Aloe dorotheae (Photographer: G.F. Smith)

Synonyms

A. harmsii A.Berger

Etymology

dorotheae: For Miss Dorothy Westhead, London.

harmsii: For Hermann A.T. Harms (1870–1942), German botanist.

Aloe downsiana T.A.McCoy & Lavranos

Etymology

downsiana: For Dr Philip E. Downs (1938–), British dentist formerly of South Africa, now based in New Zealand, for his interest in aloes and efforts to further their conservation.

Aloe droseroides Lavranos & T.A.McCoy

Etymology

droseroides: For resembling the genus *Drosera*, with rosettes of narrow leaves covered with fine white hairs, from Greek '-oides' (resembling).

Aloe droseroides (Photographer: G. Orlando)

Aloe duckeri Christian

Aloe duckeri (Photographer: S.S. Lane)

Etymology

duckeri: For H.C. Ducker, in charge of the Cotton Experiment Station, in Malawi.

Common names

chinthembwe [82] [Nyanja] [82] [Tumbuka]
khuzi [82] [Ngoni]
lichongwe [82] [Yao]

Aloe dyeri Schönland

Aloe dyeri (Photographer: SANBI, P. Joffe)

Etymology

dyeri: For Sir William T. Thiselton-Dyer (1843–1928), British botanist, director of the Royal Botanic Gardens, Kew in 1885–1905, who sent the plant from Kew to Grahamstown in 1902, where it flowered and was described by Schönland.

Common names

icena [47] [Zulu]
ilicena [47] [Zulu]

Aloe ecklonis Salm-Dyck

Aloe ecklonis (Photographer: R.R. Klopper)

Synonyms

A. agrophila Reynolds

Etymology

agrophila: For the preferred habitat in grass, from Greek 'agros' (countryside, farm), 'philos' (friend).
ecklonis: For Christian Frederick Ecklon (1795–1868), Danish chemist and botanical explorer settling at the Cape, who first sent seeds of this plant to Europe.

Common names

Ecklon's aloe [55, 65, 95] [English]
Ecklon-se-aalwyn [55, 65, 95] [Afrikaans]
grasaalwyn [55, 62, 108] [Afrikaans]
grass aloe [55, 62, 109] [English]
grootgrasaalwyn [55, 109] [Afrikaans]
hloho tsa makaka [53, 55] [Sotho, Southern]
hloho-tsa-makaka [58] [Sotho, Southern]
isiphukuthwane [55, 95] [Zulu]
isiphuthumane [55, 65, 95] [Zulu]
isipukutwane [53, 55, 101] [Zulu]
isiputumane [55] [Zulu]
isisphukuthwane [65] [Zulu]
lekhala [58] [Sotho, Southern]
lekhalana [53, 55, 58] [Sotho, Southern]

maroba-lihale [53, 55, 58, 95, 101] [Sotho, Southern]
maroba-lilale [58] [Sotho, Southern]
maroba-lithatle [55] [Sotho, Southern]
sereleli [58] [Sotho, Southern]
vlei-aalwyn [9, 55] [Afrikaans]

Aloe edentata Lavranos & Collen.

Etymology

edentata: For the unarmed leaf margin, from Latin 'e' (without), 'dentatus' (toothed).

Aloe edouardii Rebmann

Etymology

edouardii: For Mr Edouard Andriamboavonjy, driver who accompanied the author to the field and saw the plant first.

Aloe elata S.Carter & L.E.Newton

Etymology

elata: For the tall stems, from Latin 'elatus' (tall).

Aloe elegans Tod.

Synonyms

A. abyssinica A.Berger (nom. illegit.)
A. abyssinica Lam. var. *peacockii* Baker
A. aethiopica (Schweinf.) A.Berger
A. peacockii (Baker) A.Berger
A. percrassa A.Berger (nom. illegit.) var. *saganeitiana* A.Berger
A. schweinfurthii Hook.f. (nom. illegit.)
A. vera L. var. *aethiopica* Schweinf.

Etymology

abyssinica: For the occurrence in Abyssinia.
aethiopica: For the occurrence in Ethiopia.
elegans: For the elegant appearance, from Latin.
peacockii: For John T. Peacock, collector of succulent plants, Hammersmith, England.
percrassa: For the succulent leaves, from Latin 'per-' (very), 'crassus' (thick).

Aloe elegans (Photographer: S. Demissew)

saganeitiana: For the occurrence near Saganeiti, in Ethiopia.
schweinfurthii: For Dr Georg Schweinfurth (1836–1925), German botanist, geographer and explorer of northeast Africa and Arabia.
vera: The true aloe, from Latin 'vera' (in truth, real).

Aloe elegantissima T.A.McCoy & Lavranos

Etymology

elegantissima: For the elegant appearance, from Latin, superlative.

Aloe elgonica Bullock

Aloe elgonica (Photographer: SANBI, L.C. Leach)

Aloe ecklonis

Etymology

elgonica: For the occurrence on Mt Elgon, on the Kenya-Uganda border.

Aloe elkerriana Dioli & T.A.McCoy

Etymology

elkerriana: For the occurrence at El Kerre, in Ethiopia.

Aloe ellenbeckii A.Berger

Synonyms

A. dumetorum B.Mathew & Brandham

Etymology

dumetorum: For the preferred habitat in thickets, from Latin 'dumetum' (thicket).
ellenbeckii: For Dr Hans Ellenbeck (fl. 1899–1901), German physician who collected material for Berlin on Baron Von Erlanger's expedition to East Africa.

Aloe eminens Reynolds & P.R.O.Bally

Aloe eminens (Photographer: G. Orlando)

Etymology

eminens: For the conspicuousness in nature, standing out, from Latin.

Common names

daar der [55, 102] [Somali]
dacar dheer [55] [Unspecified language]

Aloe eremophila Lavranos

Aloe eremophila (Photographer: G. Orlando)

Etymology

eremophila: For the habitat preference in desert, from Greek 'eremos' (solitary, deserted), 'philos' (friend).

Aloe erensii Christian

Etymology

erensii: For Jan Erens (1911–1982), Dutch horticulturalist and collector, who emigrated to South Africa in 1914, and also collected in East Africa.

Aloe ericahenriettae T.A.McCoy

Etymology

ericahenriettae: For Erica Henrietta McCoy, daughter of the author.

Aloe ericetorum Bosser

Etymology

ericetorum: For the habitat preference in moors, from Latin 'ericetum' (heath, moor).

Aloe erinacea D.S.Hardy

Aloe erinacea (Photographer: E.J. van Jaarsveld)

Synonyms

A. melanacantha A.Berger var. *erinacea* (D.S.Hardy) G.D.Rowley

Etymology

erinacea: For the prickly appearance of the leaf rosettes, hedgehog-like, from Latin 'erinaceus' (hedgehog).

melanacantha: For the black thorns on the leaf margins, from Greek 'melas, melano-' (black), 'akantha' (thorn, spine).

Common names

krimpvarkie [55, 59, 60, 75] [Afrikaans]

Aloe erythrophylla Bosser

Aloe erythrophylla (Photographer: G. Orlando)

Etymology

erythrophylla: For the red colour of the leaf, from Greek 'erythros' (red), 'phyllon' (leaf).

Aloe esculenta L.C.Leach

Aloe esculenta (Photographer: SANBI, L.C. Leach)

Etymology

esculenta: Because there are reports that the flowers are edible[72], from Latin 'esculentus' (edible).

Common names

dishashanogha [55] [Iraqw]
ekundu [55] [Kwangali]
endobo [55] [Kwangali] [105] [Kua]
endombwe [55] [Kwangali]
lishashankogha [55] [Diriku]
omandobo [105] [Kua]

Aloe eumassawana S.Carter, M.G.Gilbert & Sebsebe

Etymology

eumassawana: For *Aloe massawana*, with which the taxon was previously confused and which, despite the name, does not come from Massawa, from Greek 'eu' (truly).

Aloe excelsa A.Berger var. *breviflora* L.C.Leach

Aloe excelsa var. *breviflora* (Photographer: S.S. Lane)

Etymology

breviflora: For the shorter flowers, from Latin 'brevis' (short), '-florus' (flowered).
excelsa: For the growth habit, from Latin 'excelsus' (tall, high).

Aloe excelsa A.Berger var. *excelsa*

Aloe excelsa var. *excelsa* (Photographer: J.J. Meyer)

Etymology

excelsa: For the growth habit, from Latin 'excelsus' (tall, high).

Common names

chigiakia [8, 129] [Tswana]
chikowa [8, 129] [Shona]
chinungu [8, 129] [Shona]
chinyangami [8, 129] [Shona]
gavakava [8, 45, 55, 129] [Shona]
godzongo [8, 129] [Shona]
gweravana [8, 129] [Shona]
icena [55] [Ndebele]
imangani [99] [Shona]
inhlaba [8, 45, 129] [Ndebele]
mhangani [8, 129] [Tswana]
mundumba [99] [Unspecified language]
Rhodesian tree aloe [55, 126] [English]
rumangamunu [99] [Shona]
rumangamuru [99] [Unspecified language]
rumhangamhuno [8, 129] [Shona]
ruvati [8, 129] [Shona]

tree aloe [8, 55] [English]
tshikhopha [57] [Venda]
wundumba [99] [Shona]
Zimbabwe aloe [9, 55, 57, 109, 123, 124] [English]
Zimbabwe tree aloe [55] [English]
Zimbabwe-aalwyn [55, 57, 109, 123, 124] [Afrikaans]

Aloe eximia Lavranos & T.A.McCoy

Etymology

eximia: For the remarkably tall habit of growth, from Latin 'eximius' (distinguished).

Aloe falcata Baker

Aloe falcata (Photographer: R. de Villiers)

Etymology

falcata: Meaning curved like a sickle, referring to the leaf shape[42, 123] or to the curvature of the peduncle[53], from Latin 'falcatus' (falcate).

Common names

Vanrhynsdorpaalwyn [9, 55] [Afrikaans]

Aloe ferox Mill.

Synonyms

A. ferox Mill. var. *erythrocarpa* A.Berger
A. ferox Mill. var. *galpinii* (Baker) Reynolds

Aloe ferox (Photographer: N.R. Crouch)

A. ferox Mill. var. *hanburyi* Baker
A. ferox Mill. var. *incurva* Baker
A. ferox Mill. var. *subferox* (Spreng.) Baker
A. galpinii Baker
A. horrida Haw.
A. muricata Haw.
A. perfoliata Thunb.
A. perfoliata L. var. γ L.
A. perfoliata L. var. ε L.
A. perfoliata L. var. ζ Willd.
A. perfoliata L. var. [θ] *ferox* (Mill.) Aiton
A. pseudo-ferox Salm-Dyck
A. subferox Spreng.
A. supralaevis Haw.
A. supralaevis Haw. var. *erythrocarpa* Baker
Pachidendron ferox (Mill.) Haw.
P. pseudo-ferox (Salm-Dyck) Haw.
P. supralaeve (Haw.) Haw.

Etymology

erythrocarpa: For the red fruit, from Greek 'erythros' (red), 'karpos' (fruit).
ferox: For the prickly leaves, from Latin 'ferox' (fierce).
galpinii: For Ernest Edward Galpin (1858–1941), South African botanist and plant collector who discovered the plant at Queenstown, in South Africa.

hanburyi: For Sir Thomas Hanbury (1832–1907) who founded the Hanbury Botanic Gardens (La Mortola) near Ventimiglia in Italy in 1867.
horrida: For the numerous spines, from Latin 'horridus' (prickly).
incurva: For the leaves curved inwards, from Latin 'incurvus' (curved inwards).
muricata: For the prickled leaves, from Latin 'muricatus' (muricate).
perfoliata: For the stem passing through the leaves, i.e. the leaves are amplexicaul, from Latin 'per' (through), 'folia' (leaf).
pseudo-ferox: For resembling *Aloe ferox*, from Greek 'pseudo-' (false).
subferox: For resembling *Aloe ferox,* from Latin 'sub' (almost).
supralaeve/supralaevis: For the smooth upper surface of the leaf, from Latin 'supra' (above), 'laevis' (smooth).

Common names

acibara [73] [Spanish]
áloe del cabo [73] [Spanish]
aloès du Cap [11, 55] [French]
bergaalwyn [9, 55, 62, 98, 108] [Afrikaans]
bitter aloe [11, 53, 55, 94, 99, 109, 122, 123, 124, 125] [English]
bitteraalwyn [9, 53, 55, 62, 81, 94, 98, 99, 108, 109, 122, 124] [Afrikaans]
Cape aloe [11, 56, 127] [English]
Cape aloe gel [122] [English]
Cape aloes (bitter fraction) [122] [English]
Cape prickly aloe [55] [English]
common aloe [55] [English]
cultivated aloe [124] [English]
goree [55] [Unspecified language]
goreebosch [108] [Afrikaans]
hlaba [53, 55, 58, 99, 125] [Sotho, Southern]
ikhala [53, 55, 94, 96, 99, 119] [Xhosa]
inhlaba [94, 96, 99] [Zulu]
Kaapse aalwyn [55] [Afrikaans]
kanniedood [99, 125] [Afrikaans]
Kapaloë [55] [German]
Karoo-aalwyn [55, 109] [Afrikaans]
kraalaalwyn [55, 108] [Afrikaans]
lekhala [58] [Sotho, Southern]
lekhala la Quthing [53, 55, 99] [Sotho, Southern]
lekhala-la-Quthing [58, 96] [Sotho, Southern]
lekhala-le-leholo [58, 96] [Sotho, Southern]
lidah buaya [99] [Unspecified language]
makaalwyn [55, 98, 108, 124] [Afrikaans]
medicinal aloe [107] [English]
mohalakane [58] [Sotho, Southern]
new aloes [55] [English]
opregte aalwyn [55, 108] [Afrikaans]
opregte-aalwyn [98] [Afrikaans]
red aloe [11, 53, 55, 99] [English]
red aloes [55] [English]
regte-aalwyn [98, 108] [Afrikaans]
Swellendam-aalwee [98] [Afrikaans]
Swellendamaalwyn [124] [Afrikaans]
Swellendamsaalwee [108] [Afrikaans]
Swellendamsaalwyn [55] [Afrikaans]
tap aloe [124] [English]
tapaalwee [55, 108] [Afrikaans]
tapaalwyn [53, 55, 81, 99, 108, 124] [Afrikaans]
tap-aalwyn [98] [Afrikaans]
umhlaba [122] [Sotho, Southern] [55, 99, 122, 125] [Xhosa] [24, 53, 55, 99, 122, 125] [Zulu] [107] [Unspecified language]
umhlakahla [55] [Zulu]
unomaweni [119] [Xhosa]

Products

Aloe: Cape aloe yielding >50% water-soluble extractive[89].
Aloe Capensis: The residue obtained by evaporating the juice of leaves, contains >18% hydroxyanthracene derivatives calculated as anhydrous barbaloin[23].
Cape aloes: Concentrated and dried juice of the leaves; >18% hydroxyanthracene derivatives expressed as barbaloin[17, 19, 127].

Aloe fibrosa Lavranos & L.E.Newton

Etymology

fibrosa: For the presence of fibres in the leaves, from Latin 'fibrosus' (fibrous).

Aloe fievetii Reynolds var. *altimatsiatrae* (J.-B.Castillon) J.-B.Castillon

Synonyms

A. altimatsiatrae J.-B.Castillon
A. estevei Rebmann

Etymology

altimatsiatrae: For the occurrence in the High Matsiatra province in Madagascar, from Latin 'altus' (high).
estevei: Dedicated to an unnamed friend who accompanied the author on his expeditions to Madagascar.
fievetii: For Gerard Fievet, French wine-grower and succulent plant enthusiast in Madagascar.

Aloe fievetii Reynolds var. *fievetii*

Aloe fievetii var. *fievetii* (Photographer: SANBI, G.W. Reynolds)

Etymology

fievetii: For Gerard Fievet, French wine-grower and succulent plant enthusiast in Madagascar.

Aloe fimbrialis S.Carter

Etymology

fimbrialis: For the fimbriate leaf margin, from Latin.

Aloe fleurentinorum Lavranos & L.E.Newton

Etymology

fleurentinorum: For Jacky and Martine Fleurentin, French medical technician and his wife, resident in Yemen.

Aloe fleuretteana Rauh & Gerold

Etymology

fleuretteana: For Mrs Fleurette Andriantsjlavo, Head of the Direction de la Planification des Eaux et des Forêts, in Madagascar.

Aloe flexilifolia Christian

Etymology

flexilifolia: For the flexible leaves, from Latin 'flexilis' (flexible), 'folius' (leaved).

Common names

luza [55] [Shambala]

Aloe florenceae Lavranos & T.A.McCoy

Etymology

florenceae: For Mrs Florence Razafindratsira, the wife of the discoverer of this species, Alfred Razafindratsira.

Aloe forbesii Balf.f.

Aloe forbesii (Photographer: G.F. Smith)

Etymology

forbesii: For Dr Henry O. Forbes (1851–1932), Scottish naturalist and plant collector.

Aloe fosteri Pillans

Aloe fosteri (Photographer: SANBI, P. Joffe)

Etymology

fosteri: For Cyril Foster, an *Aloe* enthusiast from Krugersdorp, South Africa, who collected the plant.

Common names

Foster's aloe [55] [English]
tookgo [53, 55] [Sotho, Southern]

Aloe fouriei D.S.Hardy & Glen

Etymology

fouriei: For Stephanus P. Fourie of the erstwhile Transvaal Nature Conservation Division, South Africa, who first discovered it.

Aloe fouriei (Photographer: SANBI, P. Joffe)

Aloe fragilis Lavranos & Röösli

Etymology

fragilis: For the rosettes that are easily damaged, from Latin 'fragilis' (fragile).

Aloe framesii L.Bolus

Aloe framesii (Photographer: A.W. Klopper)

Synonyms

A. amoena Pillans
A. microstigma Salm-Dyck subsp. *framesii* (L.Bolus) Glen & D.S.Hardy

Etymology

amoena: For the beauty of the plant, from Latin 'amoena' (beautiful).
framesii: For Percival ('Percy') Ross Frames (1863–1947), South African solicitor,

Aloe fosteri

collector and grower of succulents, who first collected the plants in Namaqualand.

microstigma: For the small white spots on the leaves, from Greek 'mikros' (small), 'stigma' (spot, stigma).

Common names

bitter aloe [71] [English]
bitteraalwyn [71] [Afrikaans]

Aloe francombei L.E.Newton

Etymology

francombei: For Colin Francombe, ranch manager in Kenya.

Aloe friisii Sebsebe & M.G.Gilbert

Etymology

friisii: For Prof. Ib Friis (1945–), Danish botanist at the University of Copenhagen.

Aloe fulleri Lavranos

Etymology

fulleri: For Major Andrew B.I. Fuller, plant collector in southwestern Arabia.

Aloe gariepensis (Photographer: A.W. Klopper)

Aloe gariepensis Pillans

Etymology

gariepensis: For the distribution, from the Khoi name Gariep, for the Gariep River (also known as the Orange River), in South Africa, meaning 'large, huge'.

Common names

afgeronde-aalwyn [60, 75] [Afrikaans]
Gariep aloe [25] [English]
grootrivieraalwyn [9, 55] [Afrikaans]
Keimoesaalwyn [9, 55] [Afrikaans]
Orange River aloe [55, 71] [English]
Oranjerivier-aalwyn [71] [Afrikaans]
rooi-aalwyn [60, 75] [Afrikaans]

Aloe gerstneri Reynolds

Etymology

gerstneri: For Father Jacob Gerstner (1888–1948), Bavarian Roman Catholic Missionary and botanist, first collector of this species in Zululand (KwaZulu-Natal) in 1931. He was Superior of Mission Farms in Zululand in 1928–1942.

Common names

bergaalwyn [9, 55, 99] [Afrikaans]
Gerstner's aloe [95] [English]
isihlabana [49] [Zulu]
isihlabane [53, 55, 95, 99, 123] [Zulu]

Aloe gerstneri (Photographer: SANBI, G.W. Reynolds)

Aloe gilbertii T.Reynolds ex Sebsebe & Brandham subsp. *gilbertii*

Etymology

gilbertii: For Mike G. Gilbert (1943–), English botanist resident in Ethiopia and Kenya in 1968–1982, who collected the type.

Aloe gilbertii T.Reynolds ex Sebsebe & Brandham subsp. *megalacanthoides* M.G.Gilbert & Sebsebe

Etymology

gilbertii: For Mike G. Gilbert (1943–), English botanist resident in Ethiopia and Kenya in 1968–1982, who collected the type.

megalacanthoides: For resembling *Aloe megalacantha* in habit, from Greek '-oides' (resembling).

Aloe gillettii S.Carter

Etymology

gillettii: For Jan B. Gillett (1911–1995), English botanist at the Royal Botanic Gardens, Kew, resident in Kenya in 1963–1984.

Aloe glabrescens (Reynolds & P.R.O.Bally) S.Carter & Brandham

Synonyms

A. rigens Reynolds & P.R.O.Bally var. *glabrescens* Reynolds & P.R.O.Bally

Etymology

glabrescens: For the perianth surface becoming glabrous, from Latin.
rigens: For the stiff leaves, from Latin 'rigens' (rigid).

Aloe glauca Mill.

Synonyms

A. glauca Mill. var. *elatior* Salm-Dyck
A. glauca Mill. var. *humilior* Salm-Dyck
A. glauca Mill. var. *major* Haw.
A. glauca Mill. var. *minor* Haw.
A. glauca Mill. var. *muricata* (Schult.) Baker
A. glauca Mill. var. *spinosior* Haw.
A. muricata Schult. (nom. illegit.)
A. perfoliata L. var. [ζ] *glauca* (Mill.) Aiton
A. perfoliata L. var. κ L.
A. rhodacantha DC.

Etymology

elatior: For the height, from Latin 'elatus' (tall), comparative.
glauca: For the grey-green leaf colour, from Latin 'glaucus' (glaucous).

Aloe glauca (Photographer: G.F. Smith)

humilior: For being smaller, from Latin 'humilis' (low, modest), comparative.
major: For the size, from Latin 'magnus' (great), comparative.
minor: For the smaller size, from Latin, comparative of 'parvus' (small).
muricata: For the prickled or tuberculate leaves, from Latin 'muricatus' (muricate).
perfoliata: For the stem passing through the leaves, i.e. the leaves are amplexicaul, from Latin 'per' (through), 'folia' (leaf).
rhodacantha: For the red spines on the leaf margins, from Greek 'rhodos' (rose-red), 'akanthos' (spine).
spinosior: For the larger prickles near the leaf tips, from Latin 'spinosus' (spiny), comparative.

Common names

blou aalwyn [55, 123] [Afrikaans]
blouaalwee [108] [Afrikaans]
blouaalwyn [55, 62, 108] [Afrikaans]

Aloe globuligemma Pole-Evans

Aloe globuligemma (Photographer: M.J. Kimberley)

Etymology

globuligemma: For the globular flower buds, from Latin 'globulus' (little ball), 'gemma' (bud).

Common names

Gavakava [45, 55] [Shona]
icena [45, 55] [Ndebele]
knoppies aalwyn [55, 62, 123] [Afrikaans]
knoppiesaalwyn [108] [Afrikaans]
lekopane [53, 55] [Sotho, Northern]
witchdoctor's aloe [55, 126] [English]

Aloe gneissicola (H.Perrier) J.-B.Castillon & J.-P.Castillon

Synonyms

A. capitata Baker var. *gneissicola* H.Perrier

Etymology

capitata: For the head-like inflorescence, from Latin 'capitatus' (capitate).
gneissicola: For the occurrence on gneiss rock, from German 'gneiss' and Latin '-cola' (inhabiting).

Aloe gneissicola (Photographer: SANBI, P. Joffe)

Etymology

gossweileri: For John Gossweiler (1873–1952), Swiss botanist and plant collector in Angola in 1900–1950.

Aloe gossweileri Reynolds

Aloe gossweileri (Photographer: SANBI, G.W. Reynolds)

Aloe gracilicaulis Reynolds & P.R.O.Bally

Aloe gracilicaulis (Photographer: SANBI, G.W. Reynolds)

Etymology

gracilicaulis: For the delicate stems, from Latin 'gracilis' (slender), 'caulis' (stem).

Common names

daar der [55, 102] [Somali]
dacar dheer [55] [Unspecified language]
soft distant sword-leaved aloe [55] [English]

Aloe gneissicola

Aloe graciliflora Groenew.

Aloe graciliflora (Photographer: G.F. Smith)

Etymology

graciliflora: For the longer and narrower flowers, as compared to *Aloe davyana*, from Latin 'gracilis' (slender), '-florus' (flowered).

Aloe gracilis Haw.

Synonyms

A. laxiflora N.E.Br.

Etymology

gracilis: For the slender stems, from Latin 'gracilis' (slender).

Aloe gracilis (Photographer: G.F. Smith)

laxiflora: For the lax inflorescence, from Latin 'laxus' (lax), '-florus' (flowered).

Common names

rankaalwee [108] [Afrikaans]
rankaalwyn [55, 108, 109] [Afrikaans]
scrambling aloe [55, 109] [English]

Aloe grandidentata Salm-Dyck

Etymology

grandidentata: For the large teeth on the leaf margins, from Latin 'grandis' (large), 'dentatus' (toothed), an innapropriate name as the thorns are not really larger than in other species of maculate aloe[53, 123].

Common names

aanteelaalwyn [55, 109] [Afrikaans]
bontaalwyn [55, 62, 108, 123] [Afrikaans]
kanniedood [55, 62, 108, 109, 123] [Afrikaans]
kleinbontaalwyn [55, 62, 108, 109] [Afrikaans]

Aloe grandidentata (Photographer: A.W. Klopper)

Aloe grata Reynolds

Aloe grata (Photographer: SANBI, G.W. Reynolds)

Etymology

grata: For the pleasing appearance of the plant, from Latin 'gratus' (pleasing).

Aloe greatheadii Schönland

Aloe greatheadii (Photographer: SANBI, P. Joffe)

Synonyms

A. pallidiflora A.Berger
A. termetophyla De Wild.

Etymology

greatheadii: For Dr J.B. Greathead who collected the type with Dr S. Schönland.
pallidiflora: For the pale flower, from Latin 'pallidus' (pale), '-florus' (flowered).
termetophyla: For the habitat on termite mounds, from Latin 'termes' (those that terminate, or destroy) and Greek 'phylos' (friend).

Common names

bontblaaraalwyn [99] [Afrikaans]
chigiakia [8, 129] [Tswa]
chikowa [8, 118, 129] [Shona]
chinungu [8, 118, 129] [Shona]
chinyangami [8, 129] [Tonga]
gavakava [8, 45, 55, 118, 129] [Shona]
godzongo [8, 118, 129] [Shona]
Greathead's aloe [118] [English]
Greathead's spotted leaf aloe [55, 126] [English]
Greatheads spotted-leaf aloe [7] [English]
gweravana [8, 118, 129] [Shona]
icena [8, 45, 55, 118, 129] [Ndebele]
inhlaba [118] [Ndebele]
itembushia [55, 102] [Kituba]
kizimabupia [102] [Kituba]
kizima-bupia [55, 123] [Kituba]
kleinaalwyn [99] [Afrikaans]
kxophane [99] [Tswana]
mbudyaudya [8, 129] [Shona]
mhangani [8, 129] [Tswa]
mubudyadya [118] [Shona]
rumhangamhuno [8, 129] [Shona]
ruvati [8, 118, 129] [Shona]
sekgopha [55] [Venda]

Aloe greenii Baker

Aloe greenii (Photographer: N.R. Crouch)

Etymology

greenii: There appears to be no record of the person by the name of Green commemorated here[53], C.G. or G.H. Green have been suggested[42].

Common names

icena [47, 95, 123] [Zulu]
ilicena [47] [Zulu]

Aloe grisea S.Carter & Brandham

Etymology

grisea: For the grey colour of the leaves, from Latin 'griseus' (grey).

Aloe guerrae Reynolds

Aloe guerrae (Photographer: SANBI, G.W. Reynolds)

Etymology

guerrae: For Dr Guilherme Guerra, director of Agriculture and Forests in Angola.

Common names

chaudala [55, 102] [Unspecified language]

Aloe guillaumetii Cremers

Etymology

guillaumetii: For Dr Jean L. Guillaumet (1934–), French plant ecologist at Office de la Recherche Scientifique et Technique d'Outre-Mer (ORSTOM), specializing in the vegetation of Madagascar.

Aloe haemanthifolia A.Berger & Marloth

Aloe haemanthifolia (Photographer: E.J. van Jaarsveld)

Etymology

haemanthifolia: For the leaves resembling those of *Haemanthus*, from Latin 'folius' (leaved).

Aloe haggeherensis T.A.McCoy & Lavranos

Etymology

haggeherensis: For the occurrence on the Haggeher Mountains, Socotra.

Aloe hahnii Gideon F.Sm. & Klopper

Aloe hahnii (Photographer: N. Hahn)

Etymology

hahnii: For Dr Norbert Hahn, expert on the flora of the Soutpansberg in South Africa.

Aloe ×*hanburyi* A.Borzí

Synonyms

A. ×*antoninii* A.Berger (nom. superfl.)

Etymology

antoninii: Unresolved application, for someone by the name of Antonini.

hanburyi: For Sir Thomas Hanbury (1832–1907) who founded the Hanbury Botanic Gardens (La Mortola) near Ventimiglia in Italy in 1867.

Aloe hardyi Glen

Aloe hardyi (Photographer: G.F. Smith)

Etymology

hardyi: For David S. Hardy (1931–1998), horticulturalist and former curator of the succulent plant collection at the Pretoria National Botanical Garden of SANBI, South Africa.

Aloe harlana Reynolds

Aloe harlana (Photographer: SANBI, L.C. Leach)

Etymology

harlana: For the occurrence near Harla, in Ethiopia.

Aloe haworthii Sweet

Etymology

haworthii: For Adrian H. Haworth (1768–1833), English zoologist and botanist, and succulent plant specialist.

Aloe haworthioides Baker var. *aurantiaca* H.Perrier

Etymology

aurantiaca: For the colour of the flower, from Latin 'aurantiacus' (orange).

haworthioides: For resembling representatives of the related asphodeloid genus *Haworthia*, from Greek '-oides' (similar to).

Common names

sarivahona 55 [Unspecified language]

Aloe haworthioides Baker var. *haworthioides*

Aloe haworthioides *var. haworthioides* (Photographer: SANBI, W. Rauh)

Synonyms

Leemea haworthioides (Baker) P.V.Heath

Etymology

haworthioides: For resembling representatives of the related asphodeloid genus *Haworthia*, from Greek '-oides' (similar to).

Aloe hazeliana Reynolds var. *hazeliana*

Aloe hazeliana var. *hazeliana* (Photographer: SANBI, D.C.H. Plowes)

Etymology

hazeliana: For Hazel O. Munch (née Elske) (1912–2001) who explored the Chimanimani Mountains in Zimbabwe with spouse Raymond Charles Munch (1901–1985), a farmer near Rusape, Zimbabwe. They established a garden of native plants, including aloes and cycads.

Common names

Hazel's rock aloe [55, 126] [English]

Aloe hazeliana Reynolds var. *howmanii* (Reynolds) S.Carter

Synonyms

A. howmanii Reynolds

Aloe hazeliana var. *howmanii* (Photographer: M.J. Kimberley)

Etymology

hazeliana: For Hazel O. Munch (née Elske) (1912–2001) who explored the Chimanimani Mountains in Zimbabwe with spouse Raymond Charles Munch (1901–1985), a farmer near Rusape, Zimbabwe. They established a garden of native plants, including aloes and cycads.

howmanii: For Roger Howman, Native Commissioner at Ndanga and Zaka until 1939, later at Melsetter, in Rhodesia (Zimbabwe).

Common names

Howman's cliff aloe [55, 126] [English]

Aloe helenae Danguy

Aloe helenae (Photographer: A.W. Klopper)

Etymology

helenae: For Mrs Helen Decary, wife of Raymond Decary, French financial administrator and botanist in Madagascar.

Aloe heliderana Lavranos

Etymology

heliderana: For the occurrence near Helidera, in Somalia.

Aloe hemmingii Reynolds & P.R.O.Bally

Etymology

hemmingii: For C.F. Hemming, of the Desert Locust Survey.

Aloe hemmingii (Photographer: SANBI, P.R.O. Bally)

Aloe hendrickxii Reynolds

Aloe hendrickxii (Photographer: SANBI, P. Schlieben)

Etymology

hendrickxii: For Fred L. Hendrickx, Belgian agronomist in Central Africa.

Aloe hereroensis Engl. var. *hereroensis*

Aloe hereroensis var. *hereroensis* (Photographer: S.P. Bester)

Synonyms

A. hereroensis Engl. var. *orpeniae* (Schönland) A.Berger
A. orpeniae Schönland

Etymology

hereroensis: For the occurrence in the region inhabited by the Herero tribe, in Namibia.
orpeniae: For Kate Orpen (1870–1943), South African poet and plant collector, who collected the type.

Common names

aukoreb [33, 55] [Nama]
bergaalwyn [9, 55] [Afrikaans]
deurmekaarkoppie [55, 109] [Afrikaans]
Herero aloe [55] [English]
Hereroland aloe [33, 55] [English]
Hererolandaalwyn [33, 55] [Afrikaans]
koreb [33] [Nama]
otjindombo [33, 55] [Herero]
sand aalwyn [55, 59, 62] [Afrikaans]
sand aloe [55, 109] [English]
sandaalwyn [33, 60, 108, 109, 123] [Afrikaans]
vlakte-aalwyn [55, 59, 62, 108, 123] [Afrikaans]
vlakteaalwyn [55, 60, 109] [Afrikaans]

Aloe hereroensis Engl. var. *lutea* A.Berger

Aloe hereroensis var. *lutea* (Photographer: SANBI, D.S. Hardy)

Etymology

hereroensis: For the occurrence in the region inhabited by the Herero tribe, in Namibia.

lutea: For the yellow flowers, from Latin 'luteus' (yellow).

Aloe ×*hertrichii* E.Walther

(*A. lineata* × *A. vera*)

Etymology

hertrichii: For William Hertrich (1878–1966), curator of the Huntington Botanical Gardens in the United States of America.

Aloe ×*heteracantha* Baker

(*A. arborescens* × *A. maculata*)

Synonyms

A. paradoxa A.Berger

Etymology

heteracantha: For the leaves which are unarmed when young and when mature can be unarmed or toothed, from Greek 'hetero-' (different), 'akanthos' (spine).

paradoxa: Probably for the difficulty in identification, from Latin 'paradoxus' (strange, paradoxical).

Aloe heybensis Lavranos

Etymology

heybensis: For the occurrence on Buur Heybe, in Somalia.

Aloe hildebrandtii Baker

Synonyms

A. gloveri Reynolds & P.R.O.Bally

Etymology

gloveri: For Major P.E. Glover who discovered it on Gaan Libah in Somalia, in 1944.

hildebrandtii: For Dr Johann M. Hildebrandt (1847–1881), German naturalist who travelled widely and collected in Africa and Madagascar.

Aloe hlangapies Groenew.

Aloe hlangapies (Photographer: J.E. Burrows)

Etymology

hlangapies: For the occurrence at Hlangapies (Hlangapiesberg), near Piet Retief in Mpumalanga, South Africa.

Aloe hoffmannii Lavranos

Etymology

hoffmannii: For Ralph Hoffmann (fl. 1995–2002), Swiss horticulturalist and succulent plant enthusiast based near Zürich.

Aloe hereroensis var. *lutea*

Aloe ×*hoyeri* Radl
(*A. purpurea* × *A. serrulata*)

Etymology
hoyeri: Application obscure; possibly for Thomas Hoy (?–1821), gardener at Syon House in England.

Aloe humbertii H.Perrier

Etymology
humbertii: For Prof. Henri Humbert (1887–1967), French botanist in Madagascar.

Aloe humilis (L.) Mill.

Aloe humilis (Photographer: SANBI, P. Joffe)

Synonyms
A. acuminata Haw.
A. acuminata Haw. var. *major* Salm-Dyck
A. echinata Willd.
A. echinata Willd. var. *minor* Salm-Dyck
A. humilis Ker Gawl. (nom. illegit.)
A. humilis (L.) Mill. var. *acuminata* (Haw.) Baker
A. humilis (L.) Mill. var. *candollei* Baker
A. humilis (L.) Mill. var. *echinata* (Willd.) Baker
A. humilis (L.) Mill. var. *echinata* (Willd.) Baker subvar. *minor* Salm-Dyck
A. humilis (L.) Mill. var. *incurva* Haw.
A. humilis (L.) Mill. var. *incurva* Haw. subvar. *minor* (Salm-Dyck) A.Berger
A. humilis (L.) Mill. var. *macilenta* Baker
A. humilis (L.) Mill. var. *suberecta* (Aiton) Baker
A. humilis (L.) Mill. var. *suberecta* (Aiton) Baker subvar. *semiguttata* Haw.
A. humilis (L.) Mill. var. *subtuberculata* (Haw.) Baker
A. incurva (Haw.) Haw.
A. macilenta (Baker) G.Nicholson
A. perfoliata L. var. [μ] *suberecta* Aiton
A. perfoliata L. var. [o] *humilis* L.
A. suberecta (Aiton) Haw.
A. suberecta (Aiton) Haw. var. *acuminata* Haw.
A. suberecta (Aiton) Haw. var. *semiguttata* Haw.
A. subtuberculata Haw.
A. tuberculata Haw.
A. verrucosospinosa All.
Catevala humilis (L.) Medik.

Etymology
acuminata: For the leaves, from Latin 'acuminatus' (pointed).
candollei: For Prof. Augustin P. De Candolle (1778–1841), Swiss botanist.
echinata: For the leaf prickles, from Latin 'echinatus' (prickly).
humilis: For the low-growing habit, from Latin 'humilis' (modest, low).
incurva: For the leaves curved inwards, from Latin 'incurvus' (curved inwards).
macilenta: For being lean, from Latin 'macilentus' (lean).

major: For the size, from Latin 'magnus' (great), comparative.
minor: For the smaller size, from Latin, comparative of 'parvus' (small).
perfoliata: For the stem passing through the leaves, i.e. the leaves are amplexicaul, from Latin 'per' (through), 'folia' (leaf).
semiguttata: For being 'half-warted', from Latin 'semi' (half), 'guttatus' (spotted).
suberecta: For the suberect leaves, from Latin 'sub' (almost), 'erectus' (erect).
subtuberculata: For the leaf tubercules, from Latin 'sub' (almost), 'tuberculatus' (tuberculate).
tuberculata: For the numerous tubercules on the leaf surfaces, from Latin 'tuberculatus' (tuberculate).
verrucosospinosa: For the tubercles and spines on the leaves, from Latin 'verrucosus' (warted), 'spinosus' (spiny).

Common names

dwarf hedgehog aloe [9, 55, 62, 99, 101] [English]
hedgehog aloe [55, 76] [English]
krimpvarkieaalwyn [55] [Afrikaans]
serelei [55] [Sotho, Southern]

Aloe ibitiensis H.Perrier

Synonyms

A. cremersii Lavranos
A. cyrillei J.-B.Castillon
A. itremensis Reynolds
A. saronarae Lavranos & T.A.McCoy

Etymology

cremersii: For George A. Cremers (1936–), French botanist.
cyrillei: For Cyrille Rakotonanahary, primary school teacher who made many trips looking for new species of succulents.
ibitiensis: For the occurrence on Mt Ibity, in Madagascar.
itremensis: For the occurrence on the Itremo range, in Madagascar.
saronarae: For the occurrence near Saronara, in Madagascar.

Aloe ibitiensis (Photographer: S.E. Rakotoarisoa)

Aloe ifanadianae J.-B.Castillon

Etymology

ifanadianae: For the occurrence near Ifanadiana, in Madagascar.

Aloe imalotensis Reynolds var. *imalotensis*

Synonyms

A. contigua (H.Perrier) Reynolds
A. deltoideodonta Baker var. *contigua* H.Perrier

Aloe imalotensis var. *imalotensis* (Photographer: R.E. Rakotoarisoa)

Etymology

contigua: Probably for the relationship to other taxa, from Latin 'contiguus' (adjoining, neighbouring).
deltoideodonta: For the leaf marginal teeth, from Greek 'deltoides' (delta-shaped), 'odous, odontos' (tooth).
imalotensis: For the occurrence in the Imaloto Valley, in Madagascar.

Common names

vahombato [91] [Unspecified language]
vahongarana [91] [Unspecified language]

Aloe imalotensis Reynolds var. *longeracemosa* J.-B.Castillon

Etymology

imalotensis: For the occurrence in the Imaloto Valley, in Madagascar.
longeracemosa: For the long inflorescences, from Latin 'longus' (long), 'racemus' (raceme).

Aloe imalotensis var. *imalotensis*

Aloe ×imerinensis Bosser (*A. capitata* var. *capitata* × *A. macroclada*)

Etymology

imerinensis: For the occurrence in the region inhabited by the Imerina tribe in Madagascar.

Aloe inamara L.C.Leach

Aloe inamara (Photographer: SANBI, L.C. Leach)

Etymology

inamara: Because the leaves do not taste bitter, from Latin 'amarus' (bitter), 'in' (not).

Aloe inconspicua Plowes

Aloe inconspicua (Photographer: SANBI, D.C.H. Plowes)

Etymology

inconspicua: For the size and morphology of the plant that makes it difficult to find in the field, from Latin 'inconspicua' (inconspicuous).

Aloe inermis Forssk.

Aloe inermis (Photographer: SANBI, J.J. Lavranos)

Etymology

inermis: For the entire leaf margin, from Latin 'inermis' (unarmed).

Aloe inexpectata Lavranos & T.A.McCoy

Etymology

inexpectata: Because it was found unexpectedly while the collector was looking for another species, from Latin, 'inexpectatus' (unexpected).

Aloe ×insignis N.E.Br.

(*A. hexapetala* × *A. humilis*)

Etymology

insignis: For its distinctive appearance, from Latin 'insignis' (distinguished, remarkable).

Aloe integra Reynolds

Etymology

integra: For the leaf margin, which is usually entire, even though small teeth have been reported[123], from Latin 'integer' (entire).

Aloe integra (Photographer: SANBI, G.W. Reynolds)

Aloe inyangensis Christian var. *inyangensis*

Aloe inyangensis var. *inyangensis* (Photographer: J.E. Burrows)

Etymology

inyangensis: For the occurrence on Mt Inyanga, in Zimbabwe.

Common names

Inyanga aloe [7, 55, 126] [English]

Aloe inyangensis Christian var. *kimberleyana* S.Carter

Aloe inyangensis var. *kimberleyana* (Photographer: M.J. Kimberley)

Etymology

inyangensis: For the occurrence on Mt Inyanga, in Zimbabwe.

kimberleyana: For Rose and Mike Kimberley who accompanied the author in the Eastern Highlands of Zimbabwe.

Aloe irafensis Lavranos, T.A.McCoy & Al-Gifri

Etymology

irafensis: For the occurrence on Jabal Iraf, in Yemen.

Aloe irafensis (Photographer: G. Orlando)

Aloe isaloensis H.Perrier

Etymology

isaloensis: For the occurrence on the Isalo Mountains, in Madagascar.

Aloe isaloensis (Photographer: SANBI, G.W. Reynolds)

Aloe inyangensis var. *inyangensis*

Aloe jacksonii Reynolds

Aloe jacksonii (Photographer: SANBI, G.W. Reynolds)

Etymology

jacksonii: For Mr T.H.E. Jackson, Acting Civil Affairs Officer in Ethiopia, who collected the type.

Aloe jawiyon Christie, Hannon & Oakman

Aloe jawiyon (Photographer: G. Orlando)

Etymology

jawiyon: For the Soqotri common name for the plant, 'je'awiyon'.

Common names

je'awiyon [31] [Soqotri]

Aloe jibisana L.E.Newton

Etymology

jibisana: For the occurrence on Mt Jibisa, in Kenya.

Aloe johannis J.-B.Castillon

Etymology

johannis: For John J. Lavranos (1926–) Greek insurance broker, botanist and collector of succulents throughout southern and eastern Africa, Arabia and Madagascar.

Aloe johannis-bernardii J.-P.Castillon

Etymology

johannis-bernardii: For Prof. Jean-Bernard Castillon, who described many species from Madagascar.

Aloe johannis-philippei J.-B.Castillon

Etymology

johannis-philippei: For Jean-Philippe Castillon (1965–), son of the author and French Professor of Mathematics at the University of La Reunion, who has discovered many new Madagascan aloes.

Aloe jucunda Reynolds

Aloe jucunda (Photographer: SANBI, G.W. Reynolds)

Etymology

jucunda: For the attractive appearance, from Latin 'jucundus' (nice).

Aloe juddii van Jaarsv.

Etymology

juddii: For Eric Judd, aloe artist and book author.

Aloe juvenna Brandham & S.Carter

Etymology

juvenna: Misread on the original label of a cultivated plant, labelled as a possible juvenile form, pseudo-Latin, from English 'juvenile'.

Aloe kahinii T.A.McCoy & Lavranos

Etymology

kahinii: For the President of Somaliland (Somalia), Dahir Rayale Kahin.

Aloe kamnelii van Jaarsv.

Etymology

kamnelii: For Mike Kamstra and Philip Nel, keen *Aloe* enthusiasts and intrepid explorers from the Western Cape, South Africa, who brought the plants to the author's attention.

Aloe kaokoensis van Jaarsv., Swanepoel & A.E.van Wyk

Etymology

kaokoensis: For the occurrence in the Kaokoveld, in Namibia.

Aloe johannis-bernardii

Aloe kaokoensis (Photographer: E.J. van Jaarsveld)

Aloe karasbergensis Pillans

Aloe karasbergensis (Photographer: E.J. van Jaarsveld)

Synonyms

A. striata Haw. subsp. *karasbergensis* (Pillans) Glen & D.S.Hardy

Etymology

karasbergensis: For the occurrence on the Great Karasberg, in Namibia.
striata: For the lines on the leaves, from Latin 'striatus' (striate).

Aloe ×*keayi* Reynolds
(*A. buettneri* × *A. schweinfurthii*)

Etymology

keayi: For Dr Ronald W.J. Keay (1920–1998), British botanist and forestry officer in Nigeria.

Aloe kedongensis Reynolds

Aloe kedongensis (Photographer: C.S. Björa)

Synonyms

A. nyeriensis Christian subsp. *kedongensis* (Reynolds) S.Carter

Etymology

kedongensis: For the occurrence in the Kedong Valley, in Kenya.
nyeriensis: For the occurrence at Nyeri, in Kenya.

Common names

echichuviwa [55] [Turkana]
echuchuka [55] [Turkana]
esuguroi [55] [Maasai]
harguessa [55] [Borana]
jolonji [55] [Chidigo]
kigaka [55] [Lulogooli]
kiluma [6, 55] [Kamba]
kirumu [55] [Gikuyu]
kisikmamleo [55] [Swahili]
kisimando [55] [Swahili]
kitori [55] [Kigiryama]
lineke [55] [Lulogooli]
mukumi [55] [Kiembu]
ogara [55] [Dholuo]
olkos [55] [Pökoot]
osuguroi [55] [Maasai]
osuguru [6] [Maasai]
sikorowet [55] [Pökoot]
tangaratuet [55] [Nandi]
tangaratwet [6, 55] [Kipsigis]
thugurui [55] [Gikuyu]
tolkos [55] [Pökoot]

Aloe kefaensis M.G.Gilbert & Sebsebe

Etymology

kefaensis: For the occurrence in the Kef[f]a region, in Ethiopia.

Aloe ketabrowniorum L.E.Newton

Etymology

ketabrowniorum: For Ken D.F. Brown (1957–), artist, and his wife Anne E. (née Powys) (1964–), natural history consultant, explorers and collectors in Kenya, from Latin 'et' (and).

Aloe khamiesensis Pillans

Etymology

khamiesensis: For the occurrence on the Kamiesberg in South Africa, where the plants were first collected.

Aloe khamiesensis (Photographer: A.W. Klopper)

Common names

aloë boom [99] [Afrikaans]
aloeboom [55, 62, 123] [Afrikaans]
bergaalwyn [9, 55, 124] [Afrikaans]
Kamiesberg aloe [55, 71] [English]
Kamiesberg alwyn [71] [Afrikaans]
Namakwa-aalwyn [55, 124] [Afrikaans]
Namaqua aloe [55, 124] [English]
tweederly [55, 62, 99, 123] [Afrikaans]
tweederly aloëboom [108] [Afrikaans]
wilde-aalwee [108] [Afrikaans]
wildeaalwyn [55, 123] [Afrikaans]
wilde-aalwyn [55, 62, 108, 123] [Afrikaans]

Aloe kilifiensis Christian

Etymology

kilifiensis: For the occurrence near Kilifi, in Kenya.

Common names

kisimamleao [55] [Swahili]

Aloe kniphofioides Baker

Aloe kniphofioides (Photographer: J.E. Burrows)

Synonyms

A. marshallii J.M.Wood & M.S.Evans

Etymology

kniphofioides: For resembling the genus *Kniphofia*, from Greek '-oides' (resembling).
marshallii: Unresolved application, for someone by the name of Marshall.

Common names

grasaalwyn [55, 62, 86, 95, 108] [Afrikaans]
grass aloe [55, 86, 95][English]
inhlatjana [95] [Swati]

Aloe komaggasensis Kritz. & van Jaarsv.

Aloe komaggasensis (Photographer: E.J. van Jaarsveld)

Synonyms

A. striata Haw. subsp. *komaggasensis* (Kritz. & van Jaarsv.) Glen & D.S.Hardy

Etymology

komaggasensis: For the occurrence near Komaggas, in South Africa.
striata: For the lines on the leaves, from Latin 'striatus' (striate).

Aloe komatiensis Reynolds

Etymology

komatiensis: For the occurrence at Komatipoort, in South Africa.

Aloe kouebokkeveldensis van Jaarsv. & A.B.Low

Etymology

kouebokkeveldensis: For the occurrence on the Koue Bokkeveld Mountains, in South Africa.

Aloe krapohliana Marloth

Synonyms

A. krapohliana Marloth var. *dumoulinii* Lavranos

Aloe krapohliana (Photographer: R.R. Klopper)

Etymology

dumoulinii: For Jan Dumoulin (fl. 1970s), curator of the erstwhile Hester Malan Nature Reserve (now the Goegap Nature Reserve) in the Northern Cape Province, South Africa.

krapohliana: For H.J.C. Krapohl, land surveyor in South Africa, who first collected it.

Common names

bontaalwyn [78] [Afrikaans]
Krapohl's aloe [55, 71] [English]
Krapohl-se-aalwyn [71] [Afrikaans]
miniature aloe [55] [English]
painted-leaved aloe [55] [English]

Aloe kraussii Baker

Etymology

kraussii: For Dr Ferdinand F. von Krauss (1812–1890), German scientist, director of the Stuttgart Natural History Museum, traveller and collector in South Africa.

Aloe kraussii (Photographer: G. Nichols)

Common names

broad-leaved yellow grass aloe [95] [English]
hloho tsa makaka [101] [Sotho, Southern]
hloho-tsa-makaka [55, 95] [Sotho, Southern]
isiphukhutwane [55, 95] [Zulu]
isiphuthumane [95] [Zulu]
isipukutwane [101] [Zulu]
klompiesaalwyn [108] [Afrikaans]
lekhalana [55, 101] [Sotho, Southern]
lekxalana [95] [Sotho, Southern]
maroba-lihale [55, 101] [Sotho, Southern]

Aloe kulalensis L.E.Newton & Beentje

Etymology

kulalensis: For the occurrence on Mt Kulal, in Kenya.

Aloe kwasimbana T.A.McCoy & Lavranos

Etymology

kwasimbana: To commemorate the killing of a lion by the collector, from Swahili 'kwa' (place of), 'simba' (lion).

Aloe labworana (Reynolds) S.Carter

Synonyms

A. schweinfurthii Baker var. *labworana* Reynolds

Aloe labworana (Photographer: SANBI, G.W. Reynolds)

Etymology

labworana: For the occurrence in the Labwor Hills, in Uganda.
schweinfurthii: For Dr Georg Schweinfurth (1836–1925), German botanist, geographer and explorer of northeast Africa and Arabia.

Aloe laeta A.Berger var. *laeta*

Aloe laeta var. *laeta* (Photographer: R.R. Klopper)

Etymology

laeta: For the bright crimson flowers, from Latin 'laetus' (bright).

Aloe laeta A.Berger var. *maniaensis* H.Perrier

Etymology

laeta: For the bright crimson flowers, from Latin 'laetus' (bright).
maniaensis: For the occurrence near the Mania River, in Madagascar.

Aloe lanata T.A.McCoy & Lavranos

Etymology

lanata: For the woolly flowers, from Latin 'lana' (wool).

Aloe latens T.A.McCoy & Lavranos

Etymology

latens: An allusion to these plants being hidden in densely vegetated, narrow gullies and long remaining undetected, from Latin 'latens' (concealed, hidden).

Aloe lateritia Engl. var. *graminicola* (Reynolds) S.Carter

Synonyms

A. graminicola Reynolds
A. solaiana Christian

Etymology

graminicola: For the preferred habitat in grasslands, from Latin 'graminis' (grass), '-cola' (inhabiting).
lateritia: For the dark brick red flower, from Latin.
solaiana: For the occurrence at Solai, in Kenya.

Aloe labworana

Common names

tolkos [55] [Pökoot]

Aloe lateritia Engl. var. *lateritia*

Synonyms

A. amaniensis A.Berger
A. boehmii Engl.
A. campylosiphon A.Berger

Etymology

amaniensis: For the occurrence at Amani in the Usambara Mountains, Tanzania.
boehmii: For Boehm, plant collector in Africa, who collected the type in 1882.
campylosiphon: For the shape of the corolla tube, from Greek 'campylos' (bent), 'siphon' (tube).
lateritia: For the dark brick red flower, from Latin.

Common names

bikakalubamba [84] [Unspecified language]
echichuviwa [55] [Turkana]
echuchuku [55] [Turkana]
esuguroi [55] [Maasai]
harguessa [55] [Borana]
ibugubugu [84] [Nyakyusa-Ngonde]
igikakarubaamba [84] [Rwanda]
igikakarubamba [55, 102] [Rwanda]
ingarigari [84] [Rundi]
jolonji [55] [Chidigo]
kidata [55, 102] [Joba] [84] [Unspecified language]
kigagi [55] [Ganda]
kigaka [55] [Lulogooli]
kikaka [55] [Soga]
kikakalumbamba [84] [Unspecified language]
kiluma [55] [Kamba]
kirumi [55] [Gikuyu]
kisimamleo [55] [Swahili]
kisimando [55] [Swahili]
kitembo [84] [Mbunga]
kithapa [84] [Asu]
kitori [55] [Kigiryama]
lineke [55] [Lulogooli]
liperege [84] [Pogolo]
lissigiri [84] [Hehe]
litembwetembwe [84] [Hehe]
lyusi [84] [Kinga]
mlalangao [84, 85] [Swahili]
mratune [84] [Chaga]
mukumi [55] [Kiembu]
ngaka [55, 84, 102] [Unspecified language]
ngarare [55, 84, 102] [Rundi]
ogara [55] [Dholuo]
olkos [55] [Pökoot]
sikorowet [55] [Pökoot]
sizimyamuliro [84] [Unspecified language]
subiri [55] [Swahili]
tangaratuet [55] [Nandi]
tangaratwet [55] [Kipsigis]
thugurui [55] [Gikuyu]
tolkos [55] [Pökoot]

Aloe lavranosii Reynolds

Etymology

lavranosii: For John J. Lavranos (1926–) Greek insurance broker, botanist and collector of succulents throughout southern and eastern Africa, Arabia and Madagascar.

Aloe leandrii Bosser

Etymology

leandrii: For Jacques D. Leandri (1903–1982), French botanist in Madagascar.

Aloe leedalii S.Carter

Etymology

leedalii: For G. Philip Leedal (1927–1982), British geologist and priest, working for the Geological Survey in Tanzania in 1950–1953, and from 1961 as missionary in southern Tanzania, active amateur field botanist and author of handbooks on mountain plants.

Aloe lensayuensis Lavranos & L.E.Newton

Etymology

lensayuensis: For the occurrence on the Lensayu Rocks, in Kenya.

Aloe lateritia var. *graminicola*

Aloe lepida L.C.Leach

Aloe lepida (Photographer: SANBI, L.C. Leach)

Etymology

lepida: For the nice appearance of the plants, from Latin 'lepidus' (graceful).

Aloe leptosiphon A.Berger

Synonyms

A. greenwayi Reynolds
A. kirkii Baker

Etymology

greenwayi: For Dr Percy James ('Peter') Greenway (1897–1980), botanist in Amani, Tanzania, and later in charge of the East African Herbarium, in Nairobi, Kenya.

kirkii: For Sir John Kirk (1832–1922), who sent the plant from Zanzibar to Kew where it arrived in 1881.

Aloe leptosiphon (Photographer: C.S. Björa)

leptosiphon: For the narrow perianth tube, from Greek 'leptos' (fine, delicate), 'siphon' (tube).

Common names

kikoli [55] [Zigula]

Aloe lettyae Reynolds

Aloe lettyae (Photographer: SANBI, G.W. Reynolds)

Etymology

lettyae: For Cythna L. Letty (1895–1985), renowned botanical artist for the then Botanical Research Institute (now SANBI) in Pretoria, South Africa, and field botanist, who collected it.

Aloe lindenii Lavranos

Etymology

lindenii: For Dr Seymour Linden (1921–2005), American chemist and succulent plant enthusiast.

Aloe linearifolia A.Berger

Aloe linearifolia (Photographer: N.R. Crouch)

Etymology

linearifolia: For the long, narrow leaves, from Latin 'linearis' (linear), 'folius' (-leaved).

Common names

dwarf yellow grass aloe [95] [English]
inkhuphuyana [35] [Zulu]
inkuphuyana [55, 95] [Zulu]

Aloe lineata (Aiton) Haw. var. *lineata*

Aloe lineata var. *lineata* (Photographer: G.F. Smith)

Synonyms

A. dorsalis Haw.
A. lineata (Aiton) Haw. var. *glaucescens* Haw.
A. lineata (Aiton) Haw. var. *viridis* Haw.
A. perfoliata L. var. [η] *lineata* Aiton

Etymology

dorsalis: For the spines on the dorsal side of the leaf, 'keel-spined', from Latin.
glaucescens: For the blue-green colour, from Latin 'glaucus' (glaucous), '-escens' (becoming).
lineata: For the longitudinal markings on the leaves, from Latin 'lineatus' (striped).

Aloe lettyae

perfoliata: For the stem passing through the leaves, i.e. the leaves are amplexicaul, from Latin 'per' (through), 'folia' (leaf).
viridis: For the bright green leaves, from Latin 'viridis' (green).

Common names

lined red-spined aloe [101] [English]
red spined striped aloe [101] [English]

Aloe lineata (Aiton) Haw. var. *muirii* (Marloth) Reynolds

Aloe lineata var. *muirii* (Photographer: J. Kirkel)

Synonyms

A. muirii Marloth

Etymology

lineata: For the longitudinal markings on the leaves, from Latin 'lineatus' (striped).
muirii: For Dr John Muir (1874–1947), plant (and particularly drift seed) collector in the western Cape, South Africa, who collected the type.

Aloe littoralis Baker

Aloe littoralis (Photographer: J. Kirkel)

Synonyms

A. rubrolutea Schinz.
A. schinzii Baker

Etymology

littoralis: For the coastal occurrence, from the place where it was originally found but not reflecting its actual occurrence, from Latin 'littoralis' (littoral).
rubrolutea: For the flower colour, from Latin 'rubrum' (red), 'luteus' (yellow).

schinzii: For Hans Schinz (1858–1941), Swiss botanist who collected the type.

Common names

aloe of the shore [99] [English]
ananash [55] [Bengali]
aukoreb [55] [Nama]
bergaalwyn [36, 55, 59, 60, 108, 123] [Afrikaans]
bol-seoh [55] [Panjabi]
carriapolum [55] [Tamil]
chennanayakam [55] [Malayalam]
chhotakanvar [55] [Hindi]
chinikala bunda [55] [Telugu]
chinikalabanda [55] [Telugu]
chirukattali [55] [Tamil]
chota-kunwar [55] [Deccan]
elva [55] [Hindi]
elwa [55] [Hindi] [55] [Panjabi]
gaharn [55] [Malayalam]
goresib [53, 99] [Nama]
ikshuramallika [55] [Sanskrit]
kalaboel [55] [Mahali]
kanya [55] [Sanskrit]
karia polam [55] [Tamil]
kariambolam [55] [Tamil]
kattavala [55] [Malayalam]
kikakalubamba [55] [Unspecified language]
komaree [55] [Bengali]
kumari [55] [Sanskrit]
lahani kumari [55] [Mahali]
Luanda tree aloe [55, 126] [English]
mopane aloe [55, 57, 123, 124] [English]
mopane-aalwyn [9, 55, 57, 123, 124] [Afrikaans]
musabar [55] [Kashmiri]
musambar [55] [Hindi]
musambaran [55] [Tamil]
mushabhir [55] [Hindi]
mushambaram [55] [Telugu]
n||cru [117] [Kung-Ekoka]
nahani kanvar [55] [Gujarati]
oolowaton [55] [Malayalam]
otjindombo [36, 53, 55, 99] [Herero]
peria karalai [55] [Tamil]
raktapolam [55] [Tamil]
sea-side aloe [55] [English]
siroo-luttalay [55] [Tamil]
siru karalai [55] [Tamil]
sybir [55] [Panjabi]
tshikhopha [57] [Venda]
Windhoek aloe [36] [English]
Windhoekaalwyn [9, 55, 57, 59, 60, 124] [Afrikaans]

Aloe littoralis

Aloe lolwensis L.E.Newton

Etymology

lolwensis: For the occurrence near Lake Victoria, locally called Lolwe in the Luo language.

Aloe lomatophylloides Balf.f.

Synonyms

Lomatophyllum lomatophylloides (Balf.f.) Marais

Etymology

lomatophylloides: For resembling representatives of the genus *Lomatophyllum*, now included in the synonymy of *Aloe*, from Greek '-oides' (resembling).

Common names

ananas marron [3, 12, 80] [French]

Aloe longibracteata Pole-Evans

Aloe longibracteata (Photographer: A.W. Klopper)

Etymology

longibracteata: For the long floral bracts, from Latin 'longus' (long), 'bracteatus' (bracteate).

Common names

Limpopo spotted aloe [55, 109] [English]
Limpopobontaalwyn [55, 109] [Afrikaans]

Aloe longistyla Baker

Aloe longistyla (Photographer: A.W. Klopper)

Etymology

longistyla: For the long styles, from Latin 'longus' (long), 'stylus' (style).

Common names

ifouaman [55] [Gourmanchéma]
Karoo aloe [55, 62, 98, 106] [English]
Karoo-aalwyn [55, 98, 108] [Afrikaans]
ramenas [62, 98, 106, 108, 123] [Afrikaans]

Aloe luapulana L.C.Leach

Aloe luapulana (Photographer: L.C. Leach)

Etymology

luapulana: For the occurrence in the Luapula District, in Zambia.

Aloe lucile-allorgeae Rauh

Etymology

lucile-allorgeae: For Dr Lucile Allorge (1937–), Madagascar-born French botanist at the National Museum in Paris, collector in Madagascar; and daughter of the French botanist Pierre Boiteau.

Aloe luntii Baker

Aloe luntii (Photographer: G. Orlando)

Etymology

luntii: For William Lunt (1871–1904), British gardener at the Royal Botanic Gardens, Kew, who collected plants in southern Arabia in 1893.

Aloe lutescens Groenew.

Aloe lutescens (Photographer: A.W. Klopper)

Etymology

lutescens: For the gradual change from scarlet buds to yellow open flowers, from Latin 'lutescens' (becoming yellow). In addition the leaves of this species are a more yellowish green in contrast to the glaucous leaves of its relatives *A. cryptopoda* and *A. wickensii.*

Common names

Malapati aloe [55, 126] [English]
Tshipise-aalwyn [9, 55] [Afrikaans]

Aloe macleayi Reynolds

Aloe macleayi (Photographer: SANBI, G.W. Reynolds)

Etymology

macleayi: For Prof. K.N.G. MacLeay, botanist at Khartoum University, in Sudan.

Aloe macra Haw.

Aloe macra (Photographer: J.-P. Castillon)

Synonyms

Lomatophyllum macrum (Haw.) Salm-Dyck ex Schult. & Schult.f.
Phylloma macrum (Haw.) Sweet

Etymology

macra/macrum: For being lean, from Latin 'macer' (lean).

Common names

mazambron marron [80] [French]
mazambron sauvage [80] [French]

Aloe macrocarpa Tod. subsp. *macrocarpa*

Synonyms

A. borziana A.Terracc.
A. edulis A.Chev. ex Hutch. & Dalziel
A. macrocarpa Tod. var. *major* A.Berger

Etymology

borziana: For Prof. Antonino Borzi (1852–1921), Italian botanist and director of the Botanical Garden of Palermo, in Italy.
edulis: For being edible, from Latin 'edulis' (edible).

Aloe macrocarpa subsp. *macrocarpa* (Photographer: SANBI, G.W. Reynolds)

macrocarpa: For the large fruits, from Greek 'makros' (large), 'karpos' (fruit).
major: For the size, from Latin 'magnus' (great), comparative.

Common names

ifouaman [55, 102] [German]
siniani yebo [55, 102] [Baatonum]

Aloe macrocarpa Tod. subsp. *wollastonii* (Rendle) Wabuyele

Synonyms

A. angiensis De Wild.
A. angiensis De Wild. var. *kitaliensis* Reynolds
A. bequaertii De Wild.
A. lanuriensis De Wild.
A. lateritia Engl. var. *kitaliensis* (Reynolds) Reynolds
A. wollastonii Rendle

Etymology

angiensis: For the occurrence at Hangi (Angi), Kivu, in the Democratic Republic of Congo.
bequaertii: For Joseph Bequaert (1886–1982), Belgian plant collector, mostly in the Congo, who collected the type.
kitaliensis: For the occurrence near Kitale, in Kenya.
lanuriensis: For the occurrence in Lanuri, Ruwenzori, in the Democratic Republic of Congo.
lateritia: For the dark brick red flower, from Latin.
macrocarpa: For the large fruits, from Greek 'makros' (large), 'karpos' (fruit).
wollastonii: For A.F.R. Wollaston, British botanist and collector in East Africa, who collected the type.

Common names

mlalangao [14] [Swahili]

Aloe macroclada Baker

Aloe macroclada (Photographer: S.E. Rakotoarisoa)

Etymology

macroclada: For the large size of the plants, from Greek 'makros' (large), 'klados' (shoot).

Common names

vahombe [90, 91] [Unspecified language]
vahona [90, 91] [Unspecified language]

Aloe macrosiphon Baker

Aloe macrosiphon (Photographer: SANBI. G.W. Reynolds)

Synonyms

A. compacta Reynolds
A. mwanzana Christian

Etymology

compacta: For the exceptionally compactly branched inflorescence, from Latin 'compactus' (compact).
macrosiphon: For the large flowers, from Greek 'makros' (large), 'siphon' (tube).
mwanzana: For the occurrence at Mwanza, in Tanzania.

Aloe maculata All.

Aloe maculata (Photographer: G.F. Smith)

Synonyms

A. commutata Tod.
A. commutata Tod. var. *bicolor* (Baker) A.Berger
A. disticha Mill. (nom. illegit.)
A. gasterioides Baker
A. grahamii Schönland
A. grandidentata Tod. (nom. illegit.)
A. latifolia (Haw.) Haw.
A. leptophylla N.E.Br. ex Baker
A. leptophylla N.E.Br. ex Baker var. *stenophylla* Baker
A. macracantha Baker
A. maculata All. var. *ficksburgensis* (Reynolds) Dandy
A. maculosa Lam.
A. obscura Mill.
A. perfoliata L. var. *obscura* (Mill.) Aiton
A. perfoliata L. var. θ L.
A. perfoliata L. var. λ L.
A. perfoliata L. var. μ L.
A. perfoliata L. var. [τ] *saponaria* Aiton
A. picta Thunb.
A. picta Thunb. var. *major* Willd.
A. saponaria (Aiton) Haw.
A. saponaria (Aiton) Haw. var. *brachyphylla* Baker
A. saponaria (Aiton) Haw. var. *ficksburgensis* Reynolds
A. saponaria (Aiton) Haw. var. *latifolia* Haw.
A. saponaria (Aiton) Haw. var. *obscura* (Mill.) Haw.
A. trichotoma Colla
A. tricolor Baker (nom. illegit.)
A. umbellata DC.

Etymology

bicolor: For the colours of the flower, from Latin 'bi-' (two), 'color' (colour).
brachyphylla: For the short leaves, from Greek 'brachys' (short), 'phyllon' (leaf).
commutata: Application obscure, from Latin 'commutatus' (changed, changing).
disticha: Probably for the leaf arrangement, from Latin 'distichus' (distichous, two-ranked).
ficksburgensis: For the occurrence near Ficksburg in the Free State Province, South Africa.
gasterioides: For resembling representatives of the related asphodeloid genus *Gasteria*, from Greek '-oides' (resembling).
grahamii: For F. Graham who took great interest in natural history pursuits and to whom the author owed receipt of a number of South African succulents.
grandidentata: For the large teeth on the leaf margins, from Latin 'grandis' (large), 'dentatus' (toothed).
latifolia: For the wide leaves, from Latin 'latus' (broad), '-folius' (leaved).
leptophylla: For the fine leaves, from Greek 'leptos' (fine), 'phyllon' (leaf).
macracantha: For the prominent spines, from Greek 'makros' (large), 'akanthos' (spine).

maculata: For the spotted leaves, from Latin 'maculatus' (spotted).
maculosa: For the spotted leaves, from Latin 'maculosus' (spotted).
major: For the size, from Latin 'magnus' (great), comparative.
obscura: Unresolved application, from Latin 'obscurus' (indistinct, obscure).
perfoliata: For the stem passing through the leaves, i.e. the leaves are amplexicaul, from Latin 'per' (through), 'folia' (leaf).
picta: For the spots on the leaves, from Latin 'pictus' (painted).
saponaria: Because the leaves are used to make soap, from Latin 'saponarius' (soapy).
stenophylla: For the narrow leaves, from Greek 'stenos' (narrow), 'phyllon' (leaf).
trichotoma: For division in threes, probably referring to the inflorescence or to the flowers, from Latin 'trichotomus' (trichotomous).
tricolor: For three colours, probably for the variegated flower (white, red and green), from Latin 'tri-' (three), 'color' (colour).
umbellata: For the inflorescence being more or less umbellate, from Latin 'umbella' (umbel).

Common names

áloe [73] [Spanish]
aloes mydlnicowaty [97] [Polish]
amahlala [95] [Zulu]
atzavara [73] [Spanish]
balsemera [73] [Spanish]
bontaalwee [55, 108] [Afrikaans]
bontaalwyn [55, 65, 92, 95, 108, 123] [Afrikaans]
broadleaf aloe [56] [English]
broad-leaved aloe [97] [English]
common soap aloe [53, 55, 65, 95, 123] [English]
curalotodo [73] [Spanish]
icena [47, 53, 55, 65, 95, 123] [Zulu]
ihala [49] [Zulu]
ilicena [47] [Zulu]
ilihala [49] [Zulu]
ingcelwane [55, 95, 96] [Zulu] [107] [Unspecified language]
inhlaba [55] [Swati]
inocelwane [53, 55] [Xhosa]
lekhala [58, 95, 96] [Sotho, Southern]
lekhala le thaba [53, 55] [Sotho, Southern]
lekhala-la-Lesotho [58, 96] [Sotho, Southern]
lekhala-la-thaba [58] [Sotho, Southern]
lekxala-lathaba [55] [Xhosa]
lihlala [95] [Swati]
mak bontaalwyn [55] [Afrikaans]
pitazabila [30] [Spanish]
seepaalwyn [55, 108] [Afrikaans]
siniani yebo [55] [Zulu]
soap aloe [9, 55, 56, 62, 97, 126] [English]
tati [55] [Tonga]
white spotted aloe [55, 107] [English]
zabila [30] [Spanish]

Aloe madecassa H.Perrier var. *lutea* Guillaumin

Etymology

lutea: For the yellow flowers, from Latin 'luteus' (yellow).
madecassa: Meaning native, or inhabitant of Madagascar, from French 'madécasse'.

Aloe madecassa H.Perrier var. *madecassa*

Etymology

madecassa: Meaning native, or inhabitant of Madagascar, from French 'madécasse'.

Aloe mahraensis Lavranos

Etymology

mahraensis: For the occurrence in Al-Mahra Province, in Yemen.

Aloe makayana Lavranos, Rakouth & T.A.McCoy

Etymology

makayana: For the occurrence in Makay Mountains, in Madagascar.

Aloe manandonae J.-B.Castillon & J.-P.Castillon

Etymology

manandonae: For the occurrence near Manandone, in Madagascar.

Aloe mandotoensis J.-B.Castillon

Aloe mandotoensis (Photographer: S.E. Rakotoarisoa)

Synonyms

A. fontainei Rebmann

Etymology

fontainei: For Pierre Fontaine, an amateur botanist and occasional travelling companion of the author.

mandotoensis: For the occurrence near Mandoto, in Madagascar.

Aloe marlothii A.Berger subsp. *marlothii*

Aloe marlothii subsp. *marlothii* (Photographer: G.F. Smith)

Synonyms

A. marlothii A.Berger var. *bicolor* Reynolds
A. supralaevis Haw. var. *hanburyi* Baker

Etymology

bicolor: For the change of the colours of the flower from red in bud to greenish-white, from Latin 'bi-' (two), 'color' (colour).

hanburyi: For Sir Thomas Hanbury (1832–1907) who founded the Hanbury

Botanic Gardens (La Mortola) near Ventimiglia in Italy, in 1867.
marlothii: For Prof. Hermann W. Rudolf Marloth (1855–1931), German botanist, analytical chemist and pharmacist, resident in South Africa from 1883, Professor of Chemistry at Stellenbosch University in 1889–1892.
supralaevis: For the smooth upper surface of the leaf, from Latin 'supra' (above), 'laevis' (smooth).

Common names

bergaalwyn [57, 62, 86, 94, 98, 99, 108, 109, 123, 124] [Afrikaans]
bindamutshe [74] [Venda]
bindamutsho [55] [Unspecified language]
binda-mutsho [99] [Venda]
bitteraalwyn [98] [Afrikaans]
blindamutsho [57] [Venda]
boomaalwyn [55, 62] [Afrikaans]
flat-flowered aloe [4] [English]
ikala [50] [Zulu]
ikhala [53, 55, 99, 125] [Zulu]
ilikala [50,] [Zulu]
imihlaba [55, 123] [Zulu]
inhlaba [35, 94, 99] [Zulu] [53, 55, 64, 99] [Swati]
inhlabane [53, 55] [Zulu]
kghopa [55, 123] [Sotho, Southern]
khopha [53, 99] [Venda]
makaalwyn [98] [Afrikaans]
mangana grande [4] [Portuguese]
mhanga [4] [Ronga] [99, 125] [Tsonga] [37] [Unspecified language]
mhanga yikulu [4] [Changa] [4] [Ronga]
mogopa [99, 125] [Tswana]
mokgopa [55] [Tswana]
mountain aloe [55, 57, 64, 86, 94, 99, 109, 122, 123, 124, 125] [English]
mpagana [4] [Chopi]
ngopa nara [53] [Sotho, Southern]
opregte-aalwyn [98] [Afrikaans]
regte-aalwyn [98] [Afrikaans]
snuifaalwyn [9, 55, 57, 99, 124, 125] [Afrikaans]
tap-aalwyn [98] [Afrikaans]
Transvaal aloe [57, 124] [English]
tree aloe [64] [English]
tshikhopha [57, 74] [Venda]
um-hlaba [130] [Zulu]
umhlaba [49, 53, 99, 122, 123] [Zulu]
umkala [50] [Zulu]

Aloe marlothii A.Berger subsp. *orientalis* Glen & D.S.Hardy

Etymology

marlothii: For Prof. Hermann W. Rudolf Marloth (1855–1931), German botanist, analytical chemist and pharmacist, resident in South Africa from 1883, Professor of Chemistry at Stellenbosch University in 1889–1892.
orientalis: For its distribution range to the east of the range of the typical subspecies, from Latin 'orientalis' (eastern).

Aloe massawana Reynolds subsp. *massawana*

Etymology

massawana: For the occurrence at Massawa district, in Ethiopia.

Aloe massawana Reynolds subsp. *sakoankenke* (J.-B. Castillon) J.-B.Castillon

Synonyms

A. sakoankenke J.-B.Castillon

Etymology

massawana: For the occurrence at Massawa district, in Ethiopia.
sakoankenke: For the Malagasy common name for the species, 'sakoankenke'.

Common names

sakoankenke [28] [Malagasy]

Aloe mawii Christian

Etymology

mawii: For Captain A.H. Maw, owner of the property in Malawi where the type was collected.

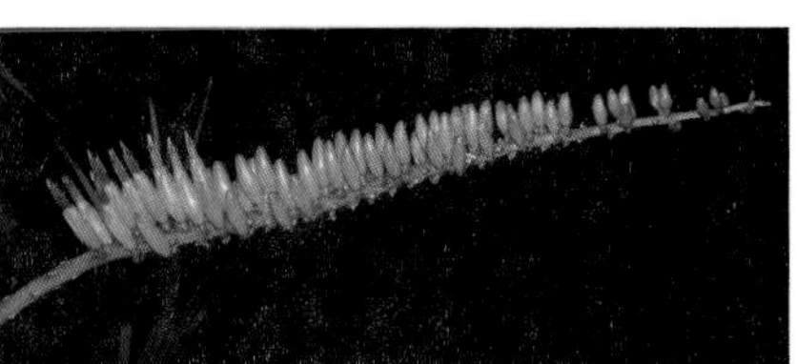

Aloe mawii (Photographer: N.R. Crouch)

Common names

chinthembwe [82] [Nyanja] [82] [Tumbuka]
khuzi [82] [Ngoni]
lichongwe [82] [Yao]

Aloe mayottensis A.Berger

Etymology

mayottensis: For the occurrence on Mayotte Island, in the Comoros Is.

Aloe mccoyi Lavranos & Mies

Aloe mccoyi (Photographer: G. Orlando)

Etymology

mccoyi: For Tom A. McCoy (1959–), American consultant and botanical collector, resident in Saudi Arabia.

Aloe mcloughlinii Christian

Etymology

mcloughlinii: For Major Alfred G. McLoughlin (1886–1960), South African lawyer and collector in northeast Africa during military service, who collected the type.

Aloe medishiana Reynolds & P.R.O.Bally

Aloe medishiana (Photographer: SANBI, G.W. Reynolds)

Etymology

medishiana: For the occurrence at Medishe, in Somalia.

Common names

daar der [55, 102] [Somali]

Aloe megalacantha Baker subsp. *alticola* M.G.Gilbert & Sebsebe

Etymology

alticola: For the occurrence at higher altitude, from Latin 'altus' (high), '-cola' (inhabiting).

megalacantha: For the large teeth on the leaf margin, from Greek 'megas, megale' (large), 'akantha' (thorn, spine).

Aloe megalacantha Baker subsp. *megalacantha*

Aloe megalacantha subsp. *megalacantha* (Photographer: SANBI, G.W. Reynolds)

Synonyms

A. magnidentata I.Verd. & Christian

Etymology

magnidentata: For the large teeth on the leaf margin, from Latin 'magnus' (large), 'dentatus' (toothed).
megalacantha: For the large teeth on the leaf margin, from Greek 'megas, megale' (large), 'akantha' (thorn, spine).

Common names

argeesaa [40] [Oromo, West Central]
argessa [40] [Wolaytta]
heejersaa [40] [Oromo, West Central]

Aloe megalocarpa Lavranos

Etymology

megalocarpa: For the large fruits, from Greek 'megas, megale' (large), 'karpos' (fruit).

Aloe melanacantha A.Berger

Aloe melanacantha (Photographer: J. Kirkel)

Etymology

melanacantha: For the black thorns on the leaf margins, from Greek 'melas, melano-' (black), 'akantha' (thorn, spine).

Common names

bittervygie [63] [Afrikaans]
black thorn aloe [55] [English]
goree [62, 108, 123] [Afrikaans]
icena [47] [Zulu]
ilicena [47] [Zulu]
klein bergaalwyn [53] [Afrikaans]
kleinbergaalwee [108] [Afrikaans]
kleinbergaalwyn [62, 71, 108, 123] [Afrikaans]
swartdoringaalwyn [9, 55] [Afrikaans]
vaalvygie [63] [Afrikaans]

Aloe ×*menachensis* (Schweinf.) Blatt. (*A. vacillans* × *A. tomentosa*)

Synonyms

A. percrassa Schweinf. var. *menachensis* Schweinf.
A. trichosantha A.Berger var. *menachensis* (Schweinf.) A.Berger

Etymology

menachensis: For the occurrence at Menacha, in Yemen.
percrassa: For the succulent leaves, from Latin 'per-' (very), 'crassus' (thick).
trichosantha: For the hairy perianth, from Greek 'trichos' (hair), 'anthos' (flower).

Aloe mendesii Reynolds

Aloe mendesii (Photographer: SANBI, L.C. Leach)

Etymology

mendesii: For Eduardo J. Mendes (1924–), Portuguese botanist who collected in Angola in the 1950s, director of the Centro de Botânica in Lisbon, Portugal.

Aloe menyharthii Baker subsp. *ensifolia* S.Carter

Etymology

ensifolia: For the ensiform leaves, from Latin 'ensis' (sword), '-folius' (leaved).
menyharthii: For Lászlò Menyárth (1849–1897), Austro-Hungarian missionary and botanist, who collected in the Zambesi region.

Aloe menyharthii Baker subsp. *menyharthii*

Aloe menyharthii subsp. *menyharthii* (Photographer: SANBI, L.C. Leach)

Etymology

menyharthii: For Lászlò Menyárth (1849–1897), Austro-Hungarian missionary and botanist, who collected in the Zambesi region.

Common names

gavi [55, 68] [Sena, Malawi]
khauzi [69] [Unspecified language]
khudsi [68] [Nyanja]

khudzi [55] [Unspecified language]
khuzi [82] [Ngoni]
mangesa [69] [Unspecified language]
manyesa [82] [Lomwe]
mdyang'oma [82] [Ngoni]
namanyesa [69] [Unspecified language]
nasi [55, 68] [Yao]
senjere [69, 82] [Unspecified language]
undyang'oma [69] [Unspecified language]
uwindi [55, 68] [Yao]

Aloe metallica Engl. & Gilg

Aloe metallica (Photographer: SANBI, L.C. Leach)

Etymology

metallica: For the metallic sheen of the leaves, from Latin 'metallicus' (metallic).

Aloe meyeri van Jaarsv.

Aloe meyeri (Photographer: E.J. van Jaarsveld)

Synonyms

A. richtersveldensis Venter & Beukes

Etymology

meyeri: For Rev. Louis G. Meyer (1867–1958), plant and insect collector and missionary in Namaqualand, who discovered the species on an expedition to the Richtersveld in 1939.
richtersveldensis: For the occurrence on the Richtersveld, in South Africa.

Aloe micracantha Haw.

Etymology

micracantha: For the small teeth on the leaf margins, although smaller teeth are known in other species[123], from Greek 'mikros' (small), 'akantha' (thorn, spine).

Common names

fynbos grass aloe [55, 109] [English]
fynbosgrasaalwyn [55, 109] [Afrikaans]
tshikhopa [74] [Venda]
tshiṱuku [74] [Venda]
wateraalwee [108] [Afrikaans]
wateraalwyn [55, 62, 108, 109] [Afrikaans]

Aloe menyharthii subsp. *menyharthii*

Aloe micracantha (Photographer: G.F. Smith)

Aloe microdonta Chiov.

Aloe microdonta (Photographer: SANBI, G.W. Reynolds)

Synonyms

A. defalcata Chiov.

Etymology

defalcata: For the curved, deflexed leaves, from Latin 'falcatus' (falcate, sickle-shaped).

microdonta: For the small teeth on the leaf margins, from Greek 'mikros' (small), 'odous, odontus' (tooth).

Common names

dacar qaraar [55] [Unspecified language]

Aloe microstigma Salm-Dyck

Aloe microstigma (Photographer: A.W. Klopper)

Synonyms

A. arabica Salm-Dyck (nom. illegit.)
A. brunnthaleri A.Berger ex Cammerl.
A. juttae Dinter

Etymology

arabica: For the occurrence in Arabia.
brunnthaleri: For Joseph Brunnthaler (1871–1914), Austrian botanist who collected in South Africa.

juttae: For Jutta Dinter (fl. 1906–1935), wife of the author.
microstigma: For the small white spots on the leaves, from Greek 'mikros' (small), 'stigma' (spot, stigma).

Common names

Karoo-aalwyn [9, 55] [Afrikaans]

Aloe millotii Reynolds

Aloe millotii (Photographer: SANBI, P. Schlieben)

Etymology

millotii: For Prof. J. Millot, French zoologist, director of the Institut de Recherche Scientifique in Madagascar, and later director of the Musée de l'Homme in Paris, France.

Aloe milne-redheadii Christian

Aloe milne-redheadii (Photographer: SANBI, L.C. Leach)

Etymology

milne-redheadii: For Edgar Milne-Redhead (1906–1996), British botanist at the Royal Botanic Gardens, Kew and field botanist in Africa.

Aloe minima Baker

Aloe minima (Photographer: J.E. Burrows)

Synonyms

A. minima Baker var. *blyderivierensis* (Groenew.) Reynolds
A. parviflora Baker
Leptaloe blyderivierensis Groenew.
L. minima (Baker) Stapf
L. parviflora (Baker) Stapf

Etymology

blyderivierensis: For the occurrence near Blyde River in Mpumalanga, South Africa.
minima: For the small size of the plant, from Latin 'minimus' (very small, smallest).
parviflora: For the small flowers, from Latin 'parvus' (small), '-florus' (flowered).

Common names

inhlatjana [95] [Swati]
isiphukhutshane [95] [Zulu]
isiphukuthwane [95] [Zulu]
isipukushane [101] [Zulu]
isipukutwane [101] [Zulu]
isiputuma [55] [Zulu]

Aloe miskatana S.Carter

Etymology

miskatana: For the occurrence at Al Miskat, in Somalia.

Aloe mitsioana J.-B.Castillon

Etymology

mitsioana: For occurrence on Mitsio Island, in Madagascar.

Aloe modesta Reynolds

Aloe modesta (Photographer: J.E. Burrows)

Etymology

modesta: For the small size of the plant, from Latin 'modestus' (modest, unassuming).

Aloe molederana Lavranos & Glen

Aloe molederana (Photographer: G. Orlando)

Etymology

molederana: For the occurrence on Molederа Hill, in Somalia.

Aloe monotropa I.Verd.

Etymology

monotropa: For the unique combination of its characters, from Greek 'monotropus' (hermit, alone and on its own).

Aloe ×*montemartinii* Catal.

Etymology

montemartinii: Unresolved application.

Aloe monticola Reynolds

Etymology

monticola: For occurring on mountains, from Latin 'mons, montis' (mountain), '-cola' (inhabiting).

Aloe minima

Aloe monticola (Photographer: S. Demissew)

Aloe morijensis S.Carter & Brandham

Etymology

morijensis: For the occurrence at Morijo, in Kenya.

Aloe mubendiensis Christian

Etymology

mubendiensis: For the occurrence at Mubende, in Uganda.

Aloe mudenensis Reynolds

Aloe mudenensis (Photographer: SANBI, G.W. Reynolds)

Etymology

mudenensis: For the occurrence in the Muden Valley in KwaZulu-Natal, South Africa.

Common names

icena [47, 95] [Zulu]
ilicena [47] [Zulu]
kleinaalwyn [55, 95, 108] [Afrikaans]
Muden aloe [95] [English]

Aloe multicolor L.E.Newton

Etymology

multicolor: For the multi-coloured perianth, from Latin 'multi-' (many), 'color' (colour).

Aloe munchii Christian

Aloe munchii (Photographer: SANBI, L.C. Leach)

Etymology

munchii: For Raymond C. Munch (1901–1985), South African farmer in Zimbabwe, with a garden containing a collection of the native flora, especially aloes and cycads.

Common names

Munch's great Chimanimani aloe [55, 126] [English]

Aloe murina L.E.Newton

Etymology

murina: For the mouse-grey colour of the inflorescences, from Latin 'murinus' (pertaining to mice).

Aloe musapana Reynolds

Aloe musapana (Photographer: SANBI, D.C.H. Plowes)

Etymology

musapana: For the occurrence at Mt Musapa, in Zimbabwe.

Common names

Musapa aloe [55, 126] [English]

Aloe mutabilis Pillans

Aloe mutabilis (Photographer: J. Kirkel)

Etymology

mutabilis: For the colour change from scarlet in bud to greenish-yellow or yellow in flower, from Latin 'mutabilis' (changeable).

Common names

blou-kransaalwyn [55, 109] [Afrikaans]
blue krantz aloe [55, 109] [English]
geelkransaalwyn [121] [Afrikaans]
kransaalwyn [9, 55, 108, 109] [Afrikaans]

Aloe mutans Reynolds

Etymology

mutans: For the colour change, the buds are red but after pollination the perianth becomes almost entirely yellow, from Latin 'mutatio' (change).

Aloe myriacantha (Haw.) Schult. & Schult.f.

Aloe myriacantha (Photographer: M.J. Kimberley)

Synonyms

A. caricina A.Berger
A. graminifolia A.Berger
A. johnstonii Baker
Bowiea myriacantha Haw.
Leptaloe caricina (A.Berger) Stapf
L. graminifolia (A.Berger) Stapf
L. johnstonii (Baker) Christian
L. myriacantha (Haw.) Stapf

Etymology

caricina: Unresolved application.
graminifolia: For the leaf shape, like a grass, from Latin 'gramen' (grass), '-folius' (leaved).
johnstonii: For Henry H. Johnston (1856–1939), British Army medical doctor and plant collector in Africa, who collected the type on Tavera, in Kenya.
myriacantha: For the many fine teeth on the leaf margins, although the spines are no more numerous than those of other grass aloes[123], from Greek 'myrios' (numerous), 'akantha' (thorn, spine).

Common names

grass aloe [55, 126] [English]
nyakaryayata [55, 102] [Nyankore]
umakhuphulwane [51, 55, 95] [Zulu]

Aloe mzimbana Christian

Aloe mzimbana (Photographer: S.S. Lane)

Etymology

mzimbana: For the occurrence at Mzimba, in Zimbabwe.

Aloe namibensis Giess

Aloe namibensis (Photographer: SANBI, D.S. Hardy)

Etymology

namibensis: For the occurrence in the Namib Desert in South-West Africa (now Namibia).

Common names

Namib aalwyn [26] [Afrikaans]
Namib aloe [26] [English]

Aloe namorokaensis (Rauh) L.E.Newton & G.D.Rowley

Synonyms

Lomatophyllum namorokaense Rauh

Etymology

namorokaense/namorokaensis: For the occurrence in the Namoroka Natural Reserve, in Madagascar.

Aloe neilcrouchii Klopper & Gideon F.Sm.

Etymology

neilcrouchii: For Prof. Neil R. Crouch of the Ethnobotany Unit of the South African National Biodiversity Institute, based at the KwaZulu-Natal Herbarium, who has added considerably to the knowledge of succulents, particularly aloes, and their uses.

Aloe neilcrouchii (Photographer: N.R. Crouch)

Aloe neoqaharensis T.A.McCoy

Etymology

neoqaharensis: For the occurrence at Jebel Qahar, Saudi Arabia, from Greek 'neos' (new), because the name *qaharensis* was already in use for another aloe.

Aloe neosteudneri Lavranos & T.A.McCoy

Etymology

neosteudneri: For its affinity to *Aloe steudneri*, named for Dr H. Steudner (1832–1863), botanist and explorer in northeast Africa, from Greek 'neos' (new).

Aloe newtonii J.-B.Castillon

Synonyms

A. intermedia (H.Perrier) Reynolds (nom. illegit.)

Etymology

intermedia: For the relationship to other taxa, from Latin 'intermedius' (intermediate).
newtonii: For Prof. Leonard E. Newton (1936–), British botanist and succulent plant expert resident in Ghana and Kenya.

Aloe ngongensis Christian

Etymology

ngongensis: For the occurrence on the Ngong Hills, in Kenya.

Aloe nicholsii Gideon F.Sm. & N.R.Crouch

Aloe nicholsii (Photographer: N.R. Crouch)

Etymology

nicholsii: For Mr Geoff Nichols who made the first known collection of this species and who pioneered the conservation-through-cultivation of many endangered medicinal and rare plants in KwaZulu-Natal, South Africa.

Aloe niebuhriana Lavranos

Etymology

niebuhriana: For Carsten Niebuhr (1733–1815), Danish mathematician and astronomer, explorer in Arabia and elsewhere.

Aloe ×nobilis Haw. (possibly *A. mitriformis* × *A. arborescens*)

Synonyms

A. mitriformis Mill. var. *spinosior* Haw.

Etymology

mitriformis: For the appearance of the rosette apex, i.e. shaped like a Bishop's cap, from Latin 'mitris' (mitre), '-formis' (shaped).
nobilis: For the size, from Latin 'nobilis' (noble).
spinosior: For being more spiny, from Latin 'spinosus' (spiny), comparative.

Aloe nordaliae Wabuyele

Etymology

nordaliae: For Prof. Inger Nordal (1944–), Norwegian botanist and plant collector in Africa.

Aloe nubigena Groenew.

Etymology

nubigena: For the high altitude distribution, meaning 'cloud-born', from Latin 'nubes' (cloud), 'genus' (birth, origin).

Common names

cloud-borne aloe [55] [English]

Aloe nubigena (Photographer: J.E. Burrows)

Aloe nuttii Baker

Aloe nuttii (Photographer: C.S. Björa)

Synonyms

A. brunneo-punctata Engl. & Gilg
A. corbisieri De Wild.
A. mketiensis Christian

Etymology

brunneo-punctata: For the spots on the leaf surface, from Latin 'brunneus' (brown), 'punctatus' (dotted).
corbisieri: For A. Corbisier-Baland (1881–?), who collected the type.
mketiensis: For the occurrence at Mketi, in Tanzania.
nuttii: For W. Harwood Nutt, missionary in Zambia in the 1890s.

Common names

dilenga [55, 102] [Unspecified language]
iwata [55, 102] [Nyamwanga]
kisimamleo [13] [Swahili]
mshubili [13] [Swahili]
msubili [13] [Swahili]
ngirya [5] [Malila]
nibeets [55, 102] [Hehe]
pomo [5] [Malila]
tembwisya [55, 102] [Mambwe-Lungu]

Aloe nyeriensis Christian

Synonyms

A. ngobitensis Reynolds

Etymology

ngobitensis: For the occurrence near Ngobit, in Kenya.
nyeriensis: For the occurrence at Nyeri, in Kenya.

Aloe occidentalis (H.Perrier) L.E.Newton & G.D.Rowley

Synonyms

Lomatophyllum occidentale H.Perrier

Aloe occidentalis (Photographer: S.E. Rakotoarisoa)

Etymology

occidentale/occidentalis: For the occurrence in western Madagascar, from Latin 'occidentalis' (western).

Aloe officinalis Forssk. var. *angustifolia* (Schweinf.) Lavranos

Synonyms

A. vera L. var. *angustifolia* Schweinf.

Etymology

angustifolia: For the narrow leaves, from Latin 'angustus' (narrow), '-folius' (leaved).
officinalis: For its medicinal use, from Latin 'officinalis' (used medicinally).
vera: The true aloe, from Latin 'vera' (in truth, real).

Aloe officinalis Forssk. var. *officinalis*

Synonyms

A. vera (L.) Burm.f. var. *officinalis* (Forssk.) Baker

Etymology

officinalis: For its medicinal use, from Latin 'officinalis' (used medicinally).
vera: The true aloe, from Latin 'vera' (in truth, real).

Aloe oligophylla Baker

Synonyms

Lomatophyllum oligophyllum (Baker) H.Perrier

Etymology

oligophylla/oligophyllum: For being few-leaved, from Greek 'oligos' (few), 'phyllon' (leaf).

Aloe omavandae van Jaarsv. & A.E.van Wyk

Aloe omavandae (Photographer: P.J.D. Winter)

Etymology

omavandae: For the occurrence at Omavanda, in Namibia.

Aloe omoana T.A.McCoy & Lavranos

Etymology

omoana: For the occurrence near the headwaters of the Omo River, in Ethiopia.

Aloe orientalis (H.Perrier) L.E.Newton & G.D.Rowley

Synonyms

Lomatophyllum orientale H.Perrier

Etymology

orientale/orientalis: For the occurrence in eastern Madagascar, from Latin 'orientalis' (eastern).

Aloe orlandi Lavranos

Aloe orlandi (Photographer: G. Orlando)

Etymology

orlandi: For Giuseppe Orlando, from Tenerife, Canary Islands, who collected the plants in Somaliland (Somalia) in 2003.

Aloe ortholopha Christian & Milne-Redh.

Etymology

ortholopha: For the row of secund flowers, from Greek 'orthos' (erect, straight), 'lophos' (crest).

Aloe ortholopha (Photographer: M.J. Kimberley)

Common names

dyke aloe [55, 126] [English]
gavakava [45, 55] [Shona]
icena [45, 55] [Ndebele]

Aloe otallensis Baker

Synonyms

A. boranensis Cufod.

Etymology

boranensis: For the occurrence in the country of the Borana, in southern Ethiopia and northern Kenya.
otallensis: For the occurrence at Otallo, in Ethiopia.

Aloe pachydactylos T.A.McCoy & Lavranos

Etymology

pachydactylos: For the extreme thickness of its relatively short leaves, from Greek 'pachys' (thick), 'dactylos' (finger).

Aloe pachygaster Dinter

Etymology

pachygaster: For the flower shape, ventricose, from Greek 'pachys' (thick), 'gaster' (stomach).

Aloe pachygaster (Photographer: E.J. van Jaarsveld)

Common names

kraalaalwyn [9, 55] [Afrikaans]

Aloe paedogona A.Berger

Aloe paedogona (Photographer: SANBI, G.W. Reynolds)

Etymology

paedogona: Because the plants are grown in Angola as a fertility charm, from Greek 'paedo-' (pertaining to children), '-gonos' (seed).

Common names

kikalangu-kibela [54, 55] [Kimbundu]

Aloe palmiformis Baker

Aloe palmiformis (Photographer: G.F. Smith)

Etymology

palmiformis: For having the form of a palm, from Latin 'palma' (palm), '-formis' (shaped).

Common names

okandole [54, 55] [Umbundu]

Aloe ×panormitana Guillaumin (hybrid with *A. brevifolia* as one parent)

Etymology

panormitana: For originating from Palermo, in Italy.

Aloe parallelifolia H.Perrier

Etymology

parallelifolia: For the strap-shaped leaves with parallel margins, from Latin 'parallelus' (parallel), '-folius' (leaved).

Aloe parvibracteata Schönland

Synonyms

A. affinis Pole-Evans (nom. illegit.)
A. decurvidens Groenew.
A. keithii Reynolds
A. lusitanica Groenew.
A. parvibracteata Schönland var. *zuluensis* (Reynolds) Reynolds
A. pongolensis Reynolds
A. pongolensis Reynolds var. *zuluensis* Reynolds

Etymology

affinis: Unresolved application, from Latin 'affinis' (allied to), possibly because it shows affinities to several maculate aloes.
decurvidens: For the decurved leaf marginal teeth, from Latin 'curvus' (curved), 'dens' (tooth).
keithii: For Captain D.R. Keith, a keen gardener in Swaziland, who brought it to the attention of the author.
lusitanica: For the occurrence in Portuguese East Africa (Mozambique), from Latin 'Lusitania' (Portugal).
parvibracteata: For the small bracts, a misleading name as the bracts are not particularly small, from Latin 'parvus' (small), 'bracteatus' (bracteate).
pongolensis: For the occurrence at Pongola, in South Africa.
zuluensis: For the occurrence in Zululand (KwaZulu-Natal), South Africa.

Common names

icena [47, 95] [Zulu]
ilicena [47] [Zulu]
imanga [53, 55] [Ronga]
inhlaba [95] [Swati]
inkalane [95] [Zulu]
Lowveld spotted aloe [55] [English]
Lowveld spotted leaf aloe [126] [English]
manga [37] [Unspecified language]
mhanga [4] [Ronga]
persbontaalwyn [55, 109] [Afrikaans]
purple spotted aloe [55, 109] [English]
spotted aloe [122] [English]
xitretre [4] [Ronga]

Aloe parvidens M.G.Gilbert & Sebsebe

Etymology

parvidens: For the small leaf teeth, from Latin 'parvus' (small), 'dens' (tooth).

Common names

daar [55] [Unspecified language]
dacar [55] [Unspecified language]

Aloe parvula A.Berger

Synonyms

A. sempervivoides H.Perrier
Leemea parvula (A.Berger) P.V.Heath

Aloe parvibracteata (Photographer: M.J. Kimberley)

Aloe parvibracteata

Aloe parvula (Photographer: S.E. Rakotoarisoa)

Etymology

parvula: For the small stature of the plants, from Latin 'parvus' (small).
sempervivoides: For resembling representatives of the genus *Sempervivum*, from Greek '-oides' (resembling).

Aloe patersonii Mathew

Etymology

patersonii: For a Mr Andrew Paterson, without further data.

Aloe pavelkae van Jaarsv., Swanepoel, A.E.van Wyk & Lavranos

Aloe pavelkae (Photographer: E.J. van Jaarsveld)

Etymology

pavelkae: For Mr Petr Pavelka, Czech succulent specialist who discovered it in Namibia.

Aloe pearsonii Schönland

Aloe pearsonii (Photographer: A.W. Klopper)

Aloe parvula

Etymology

pearsonii: For Prof. Henry Harold W. Pearson (1870–1916), English botanist and the first director of the Kirstenbosch National Botanical Garden, South Africa, who collected the type.

Common names

Pearson's aloe [36, 55, 71, 109] [English]
Pearson-se-aalwyn [71] [Afrikaans]
struikaalwyn [55, 109] [Afrikaans]

Aloe peckii P.R.O.Bally & I.Verd.

Aloe peckii (Photographer: SANBI)

Etymology

peckii: For Major E.A. Peck, officer in charge of the Veterinary and Agricultural Services in northern Somalia before and after World War II, and keen collector of the native flora.

Common names

daar [55, 102] [Unspecified language]

Aloe peglerae Schönland

Etymology

peglerae: For Alice M. Pegler (1861–1929), teacher, botanist and naturalist, plant and insect collector in South Africa.

Aloe peglerae (Photographer: J. Kirkel)

Common names

bergaalwyn [55, 62, 108, 121, 123] [Afrikaans]
grootaalwyn [55, 108] [Afrikaans]
red-hot poker [108, 123] [English]
Turk's cap aloe [9, 55] [English]
vuurpylaalwyn [55, 62] [Afrikaans]

Aloe pembana L.E.Newton

Synonyms

Lomatophyllum pembanum (L.E.Newton) Rauh

Etymology

pembana/pembanum: For the occurrence on Pemba Island, off the coast of Tanzania.

Aloe pendens Forssk.

Synonyms

A. arabica Lam.
A. dependens Steud.

Etymology

arabica: For the occurrence in Arabia.
dependens: For the hanging growth form, from Latin 'pendens' (hanging).
pendens: For the hanging growth form, from Latin 'pendens' (hanging).

Aloe penduliflora Baker

Etymology

penduliflora: For the hanging flowers, from Latin 'pendulus' (hanging down), '-florus' (flowered).

Aloe percrassa Tod.

Aloe percrassa (Photographer: S. Demissew)

Synonyms

A. abyssinica Lam. var. *percrassa* (Tod.) Baker (nom. illegit.)
A. bakeri Hook.f. ex Baker (nom. illegit.)
A. oligospila Baker
A. schimperi Schweinf (nom. illegit.)
A. schimperi G.Karst. & Schenck (nom. illegit.)

Etymology

abyssinica: For the occurrence in Abyssinia.
bakeri: For John G. Baker (1834–1920), British botanist at the Royal Botanic Gardens, Kew.
oligospila: For being few-haired, from Greek 'oligos' (few) and Latin 'pilus' (hair).
percrassa: For the succulent leaves, from Latin 'per-' (very), 'crassus' (thick).
schimperi: For Georg W. Schimper (1804–1878), German botanist and plant collector, who lived and became nationalised in Abyssinia.

Common names

arret [40, 55] [Tigrigna] [102] [Unspecified language]
daar [40] [Somali]
erreh [55, 102] [Unspecified language]
iret [55, 102] [Tigrigna]
iret matiso [40] [Tigrigna]
're harmaz [40] [Tigrigna]
zaber [40] [Tigrigna]

Aloe perdita Ellert

Etymology

perdita: For the fact that the information on the type locality was lost, and precise origin of the type collections is unknown, from Latin 'perdita' (lost).

Aloe perfoliata L.

Synonyms

A. albispina Haw.
A. commelinii Willd.
A. flavispina Haw.
A. mitraeformis Willd. (nom. illegit.)
A. mitraeformis Salm-Dyck
A. mitraeformis DC. (nom. illegit.)
A. mitriformis Mill.

Aloe perfoliata (Photographer: A.W. Klopper)

A. mitriformis Mill. var. *albispina* (Haw.) A.Berger
A. mitriformis Mill. var. *commelinii* (Willd.) Baker
A. mitriformis Mill. var. *elatior* Haw.
A. mitriformis Mill. var. *flavispina* (Haw.) Baker
A. mitriformis Mill. var. *humilior* Haw.
A. mitriformis Mill. var. *pachyphylla* Baker
A. mitriformis Mill. var. *spinulosa* (Salm-Dyck) Baker
A. mitriformis Mill. var. *xanthacantha* (Willd.) Baker
A. parvispina Schönland
A. perfoliata L. var. *mitriformis* (Mill.) Aiton
A. perfoliata L. var. κ Willd.
A. perfoliata L. var. ν L.
A. perfoliata L. var. ξ Willd.
A. spinulosa Salm-Dyck
A. xanthacantha Salm-Dyck (nom. illegit.)
A. xanthacantha Willd.

Etymology

albispina: For the white spines, from Latin 'albus' (white), 'spina' (spine).
commelinii: For either Jan or Caspar Commelin, probably Jan Commelin (1629–1692), who first studied the plants that were sent from the Cape Colony (now part of South Africa) to Amsterdam.
elatior: For the taller stems, from Latin 'elatus' (tall), comparative.
flavispina: For the yellow spines, from Latin 'flavus' (yellow), '-spinus' (spined).
humilior: For being smaller, from Latin 'humilis' (low, modest), comparative.
mitraeformis/mitriformis: For the appearance of the rosette apex, i.e. shaped like a Bishop's cap, from Latin 'mitris' (mitre), '-formis' (shaped).
pachyphylla: For the thick leaves, from Greek 'pachys' (thick), 'phyllon' (leaf).
parvispina: For the small spines, from Latin 'parvus' (small), 'spina' (spine).
perfoliata: For the stem passing through the leaves, i.e. the leaves are amplexicaul, from Latin 'per' (through), 'folia' (leaf).
spinulosa: For the very small spines, from Latin 'spinula' (small spine).
xanthacantha: For the yellow spines, from Greek 'xanthos' (yellow), 'akanthos' (spine).

Common names

kransaalwee [108] [Afrikaans]
kransaalwyn [62, 77, 78, 108, 120, 123] [Afrikaans]
mitre aloe [9, 55, 62, 108, 120, 123] [English]

Aloe perrieri Reynolds

Etymology

perrieri: For J.M. Henri A. Perrier de la Bâthie (1873–1958), French botanist, who lived in Madagascar in 1896–1933.

Aloe perryi Baker

Aloe perryi (Photographer: G. Orlando)

Etymology

perryi: For Wykeham Perry who collected plants in Socotra in 1880.

Common names

saber socotri [55] [Arabic]
sabir suqutri [55] [Arabic]
Socotra aloe [55] [English]
Socotrine aloe [56] [English]
tayf za'alhil [55] [Unspecified language]
tyf za'abeb [55] [Unspecified language]
Zanzibar aloe [56] [English]

Products

Socotrine aloes: Prepared from the juice of leaves[92].

Aloe petricola Pole-Evans

Aloe petricola (Photographer: SANBI, L.C. Leach)

Etymology

petricola: For the habitat in rocky places (granite outcrops in this case), from Latin 'petra' (rock), '-cola' (inhabiting).

Common names

rock aloe [86] [English]
rotsaalwyn [86] [Afrikaans]

Aloe petrophila Pillans

Aloe petrophila (Photographer: SANBI, G.W. Reynolds)

Etymology

petrophila: For the habitat in rocky places (cliff faces in this case), from Greek 'petra' (rock), '-philos' (friend).

Aloe peyrierasii Cremers

Synonyms

Lomatophyllum peyrierasii (Cremers) Rauh

Etymology

peyrierasii: For A. Peyrieras, French zoologist.

Aloe peyrierasii (Photographer: S.E. Rakotoarisoa)

Aloe pictifolia D.S.Hardy

Aloe pictifolia (Photographer: G.F. Smith)

Etymology

pictifolia: For the small spots on the leaves, from Latin 'pictus' (painted), '-folius' (leaved).

Aloe ×*philippei* J.-B.Castillon (*A. acutissima* var. *fiherenensis* × *A. viguieri*)

Etymology

philippei: For Jean-Philippe Castillon (1965–), son of the author and French Professor of Mathematics at the University of La Reunion, who has discovered many new Madagascan aloes.

Aloe pienaarii Pole-Evans

Etymology

pienaarii: For P.J. Pienaar, who collected it in 1914, near Pietersburg, in South Africa.

Common names

geelaalwee [108] [Afrikaans]
geelaalwyn [108] [Afrikaans]

Aloe pienaarii (Photographer: G.F. Smith)

Aloe pillansii (Photographer: H.M. Steyn)

Aloe pillansii L.Guthrie

Synonyms

A. dichotoma Masson subsp. *pillansii* (L.Guthrie) Zonn.

Etymology

dichotoma: For the branching of the stems, from Latin 'dichotomus' (dichotomous, division in pairs, forked).

pillansii: For Neville S. Pillans (1884–1964), South African botanist, who first collected it in 1926, in the Richtersveld.

Common names

bastard quiver tree [55, 124] [English]
basterkokerboom [55, 71, 124] [Afrikaans]
boom aalwyn [9, 55] [Afrikaans]
die lange [53, 55] [Afrikaans]
giant quiver tree [55, 123, 124] [English]
giant quiver-tree [36] [English]
Pillans' aloe [53, 55, 99] [English]
reusekokerboom [36, 55, 123, 124] [Afrikaans]

Aloe pirottae A.Berger

Etymology

pirottae: For Pietro R. Pirotta (1853–1936), Italian botanist and director of Rome Botanical Garden until 1928.

Common names

daar [40] [Somali]
uolaganti [40] [Aari]

Aloe pirottae (Photographer: SANBI, L.C. Leach)

Aloe plicatilis (Photographer: A.W. Klopper)

Aloe plicatilis (L.) Mill.

Synonyms

A. disticha L. var. *plicatilis* L.
A. flabelliformis Salisb.
A. lingua Thunb.
A. linguaeformis L.f. (nom. illegit.)
A. plicatilis (L.) Mill. var. *major* Salm-Dyck
A. tripetala Medik.
Kumara disticha Medik.
Rhipidodendron distichum (Medik.) Willd.
R. plicatile (L.) Haw.

Etymology

disticha/distichum: Probably for the leaf arrangement, from Latin 'distichus' (distichous, two-ranked).
flabelliformis: For the fan shape, from Latin 'flabellum' (fan), '-formis' (shaped).
lingua: For the shape of the leaves, from Latin 'lingua' (tongue).
linguaeformis: For the shape of the leaves, from Latin 'lingua' (tongue), '-formis' (shaped).
major: For the size, from Latin 'magnus' (great), comparative.
plicatile/plicatilis: Referring somewhat peripherally to the fan-shaped rosettes, from Latin, 'plicatilis' (foldable).
tripetala: Probably for the three outer perianth segments, from Greek 'tri-' (three), 'petalon' (petal).

Common names

bergaalwyn [62, 99, 108, 124] [Afrikaans]
fan aloe [9, 53, 62, 99, 109, 123, 124] [English]
Franschhoek aloe [99] [English]
Franschhoekaalwee [99, 108] [Afrikaans]
Franschhoekaalwyn [99, 109, 124] [Afrikaans]
Franschoekaalwyn [9, 53, 62] [Afrikaans]
French hoek aloe [53, 99] [English]
kokerboom [108] [Afrikaans]
tongaalwyn [99] [Afrikaans]
waaier-aalwyn [53, 55, 123] [Afrikaans]
waaieraalwyn [99, 109, 123, 124] [Afrikaans]

Aloe plowesii Reynolds

Aloe plowesii (Photographer: SANBI, D.C.H. Plowes)

Etymology

plowesii: For Darrel C.H. Plowes (1925–), South African agricultural officer and naturalist in Zimbabwe, who discovered the species.

Common names

Plowes' grass aloe [55, 126] [English]

Aloe pluridens Haw.

Synonyms

A. atherstonei Baker
A. pluridens Haw. var. *beckeri* Schönland

Aloe pluridens (Photographer: G. Nichols)

Etymology

atherstonei: For Dr William G. Atherstone (1814–1898), English medical practitioner and naturalist in South Africa.
beckeri: For Hermann F. Becker (1838–1917), plant collector from Grahamstown, in South Africa.
pluridens: Referring to the many teeth on the leaf margin, from Latin 'pluri' (many), 'dens' (teeth).

Common names

Fransaalwee [99, 108] [Afrikaans]
Fransaalwyn [55, 62, 94, 99, 108, 109, 123, 124] [Afrikaans]
French aloe [62, 94, 99, 108, 109, 123, 124] [English]
garaa [108, 124] [Afrikaans] [53] [Nama] [55, 62, 99] [Unspecified language]
many-toothed aloe [99] [English]
many-toothed tree aloe [55] [English]

Aloe polyphylla Schönland ex Pillans

Aloe polyphylla (Photographer: G.F. Smith)

Etymology

polyphylla: For the many leaves, from Greek 'poly' (many), 'phyllon' (leaf).

Common names

Basotoland aloe [55] [English]
coiled aloe [96, 101] [English]
kharatsa [58, 96] [Sotho, Southern]
kroonaalwee [108] [Afrikaans]
kroonaalwyn [62, 96, 108, 123] [Afrikaans]
lekhala [55] [Sotho, Southern]
lekhala kharatsa [53, 55] [Sotho, Southern]
lekhala-la-thaba [58, 96] [Sotho, Southern]
phurumela [55] [Sotho, Southern]
spiraalaalwyn [55] [Afrikaans]
spiral aloe [9, 55, 62, 101] [English]

Aloe porphyrostachys Lavranos & Collen. subsp. *koenenii* (Lavranos & Kerstin Koch) Lodé

Synonyms

A. koenenii Lavranos & Kerstin Koch

Etymology

koenenii: For Manfred Koenen, German horticulturalist, who collected the type in Jordan.

porphyrostachys: For the red inflorescences, from Greek 'porphyreos' (purplish-red), 'stachys' (spike).

Aloe porphyrostachys Lavranos & Collen. subsp. *porphyrostachys*

Etymology

porphyrostachys: For the red inflorescences, from Greek 'porphyreos' (purplish-red), 'stachys' (spike).

Aloe powysiorum L.E.Newton & Beentje

Etymology

powysiorum: For J. Gilfred L. Powys (1938–) and wife Patricia G. Powys, farmers, explorers and collectors of succulents in Kenya, Tanzania, southern Ethiopia and southern Sudan.

Aloe praetermissa T.A.McCoy & Lavranos

Etymology

praetermissa: Because the taxon was previously overlooked, from Latin 'praetermissus' (overlooked, missed out).

Aloe pratensis Baker

Etymology

pratensis: The name means "growing in meadows" but the species also occurs in rocky places, from Latin 'pratensis' (growing in meadows).

Common names

bergaalwyn [9, 55] [Afrikaans]
lekhala qhalane [55, 95] [Sotho, Southern]
lekhala qhalene [101] [Sotho, Southern]

Aloe pratensis (Photographer: G.F. Smith)

lekhala-la-Linakeng [96] [Sotho, Southern]
lekhala-qhalane [58, 96] [Sotho, Southern]
meadow aloe [55, 65, 95] [English]

Aloe pretoriensis Pole-Evans

Etymology

pretoriensis: For the occurrence near Pretoria, in South Africa, where the type was collected.

Common names

icena [47] [Zulu]
ilicena [47] [Zulu]
mist belt tree aloe [55, 126] [English]
Pretoria aloe [55, 109] [English]
Pretoria-aalwyn [55, 109] [Afrikaans]

Aloe pretoriensis (Photographer: A.W. Klopper)

Aloe ×*principis* (Haw.) Stearn (*A. arborescens* × *A. ferox*)

Synonyms

A. ×*caesia* Salm-Dyck
A. ×*salm-dyckiana* Schult. & Schult.f.
Pachidendron principis Haw.

Etymology

caesia: For the blue-grey colour of the leaves, from Latin 'caesius' (light blue).
principis: In appearance a princely plant, named for the Prince (Fürst) Joseph

Salm-Reifferscheid-Dyck (1773–1861), German botanist, artist, horticulturalist and succulent plant expert.
salm-dyckiana: For Prince (Fürst) Joseph Salm-Reifferscheid-Dyck (1773–1861), German botanist, artist, horticulturalist and succulent plant expert.

Common names

basteraalwyn [109] [Afrikaans]
Mossel Bay hybrid aloe [109] [English]

Aloe prinslooi I.Verd. & D.S.Hardy

Aloe prinslooi (Photographer: E.J. van Jaarsveld)

Etymology

prinslooi: For Gerry J. Prinsloo, an amateur grower of aloes, who discovered the plant.

Common names

bontaalwyn [55] [Afrikaans]
spotted aloe [55] [English]

Aloe procera L.C.Leach

Etymology

procera: For the tall inflorescences, from Latin 'procerus' (tall).

Aloe pronkii Lavranos, Rakouth & T.A.McCoy

Aloe pronkii (Photographer: A.W. Klopper)

Etymology

pronkii: For Mr Olaf Pronk of Antananarivo, Madagascar, who operates a nursery.

Aloe propagulifera (Rauh & Razaf.) L.E.Newton & G.D.Rowley

Synonyms

Lomatophyllum propaguliferum Rauh & Razaf.

Etymology

propagulifera/propaguliferum: For the production of bulbils on the inflorescence, from Latin 'propagulum' (bulbil), '-fer' (carrying).

Aloe prostrata (H.Perrier) L.E.Newton & G.D.Rowley

Synonyms

A. ankaranensis Rauh & Mangelsdorff
A. zombitsiensis Rauh & M.Teissier
Lomatophyllum prostratum H.Perrier

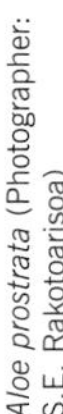

Aloe prostrata (Photographer: S.E. Rakotoarisoa)

Etymology

ankaranensis: For the occurence in the Falaise d'Ankarana, in Madagascar.
prostrata/*prostratum*: For the spreading leaves, from Latin 'prostratus' (prostrate, creeping).
zombitsiensis: For the occurrence in the Zombitsy Forest in Toliara Province, Madagascar.

Aloe pruinosa Reynolds

Aloe pruinosa (Photographer: SANBI, L.C. Leach)

Etymology

pruinosa: For the peduncle and flowers being covered in a white, waxy, powdery bloom, from Latin 'pruinosus' (covered with a waxy bloom).

Common names

icena [113] [Zulu]
icena elikhulu [113, 123] [Zulu]
kleinaalwyn [55, 62, 108, 109, 123] [Afrikaans]
poeieraalwyn [55, 109] [Afrikaans]
powder aloe [55, 109] [English]
slangkop [55, 62, 123] [Afrikaans]
slangkopaalwyn [108, 109] [Afrikaans]

Aloe pseudoparvula J.-B.Castillon

Etymology

pseudoparvula: For the resemblance to *Aloe parvula*, from Greek 'pseudo-' (false).

Aloe pseudorubroviolacea Lavranos & Collen.

Etymology

pseudorubroviolacea: For the resemblance to *Aloe rubroviolacea*, from Greek 'pseudo' (false).

Aloe ×*puberula* (Schweinf.) A.Berger (*A. camperi* × *A. trichosantha*)

Synonyms

A. vera L. var. *puberula* Schweinf.

Etymology

puberula: For the puberulous bracts, pedicels and perianth base, from Latin 'puberulus' (puberulous).
vera: The true aloe, from Latin 'vera' (in truth, real).

Aloe pubescens Reynolds

Etymology

pubescens: For the hairy flowers, from Latin 'pubescens' (pubescent).

Aloe pulcherrima M.G.Gilbert & Sebsebe

Etymology

pulcherrima: For its beauty, from Latin 'pulcher' (beautiful).

Common names

msht `iegedel fuga [40] [Guragigna]
set `ret [40] [Amharic]

Aloe purpurea Lam.

Aloe purpurea (Photographer: J.-P. Castillon)

Synonyms

A. marginalis DC. (nom. superfl.)
A. marginata (Aiton) Willd. (nom. illegit.)
A. rufocincta Haw.
Lomatophyllum aloiflorum (Ker Gawl.) G.Nicholson
L. borbonicum Willd. (nom. superfl.)
L. purpureum (Lam.) T.Durand & Schinz
L. rufocinctum (Haw.) Salm-Dyck ex Schult. & Schult.f.
Phylloma aloiflorum Ker Gawl. (nom. illegit.)
P. rufocinctum (Haw.) Sweet

Etymology

aloiflorum: For the flowers resembling those of *Aloe*, from Latin 'florus' (flowered).
borbonicum: For the occurence in the island of Bourbon (Réunion).
marginalis: For the red edge of the leaves, from Latin 'marginalis' (marginal).
marginata: For the edge of the leaves, from Latin 'marginatus' (marginate).
purpurea/purpureum: For the purple margins of the leaves, from Latin 'purpureus' (purple).
rufocincta/rufocinctum: For the rosy-edged leaves, from Latin 'rufus' (reddish), 'cinctus' (encircled).

Common names

mazambron marron [80] [French]
mazambron sauvage [80] [French]
Socotrine du pays [80] [French]

Aloe pustuligemma L.E.Newton

Etymology

pustuligemma: For the blistered surface of the flower buds, from Latin 'pustula' (blister), 'gemma' (bud).

Aloe rabaiensis Rendle

Etymology

rabaiensis: For the occurrence on the Rabai Hills, in Kenya.

Common names

folonji [55] [Chidigo]
hargeisa [55, 84] [Unspecified language]
jolonji [55, 84] [Chidigo]
kizimlo [55, 84] [Unspecified language]

Aloe ramosissima Pillans

Aloe ramosissima (Photographer: A.W. Klopper)

Synonyms

A. dichotoma Masson var. *ramosissima* (Pillans) Glen & D.S.Hardy
A. dichotoma Masson subsp. *ramosissima* (Pillans) Zonn.

Etymology

dichotoma: For the branching of the stems, from Latin 'dichotomus' (dichotomous, division in pairs, forked).
ramosissima: For being much-branched, from Latin 'ramosus' (branched), superlative.

Common names

boskokerboom [36, 59, 60, 75] [Afrikaans]
bush quiver tree [60, 75] [English]
maiden quiver tree [109] [English]
maiden's quiver tree [55, 60, 75, 123, 124] [English]
maiden's quiver-tree [36] [English]
nooienskokerboom [9, 36, 55, 60, 75, 109, 123, 124] [Afrikaans]
very much branched aloe [99] [English]

Aloe rapanarivoi J.-P.Castillon

Etymology

rapanarivoi: For S.H.J. Rapanarivo, head of the Department Flora at the Botanical and Zoological Park of Tsimbazaza in Antananarivo, Madagascar.

Aloe rauhii Reynolds

Aloe rauhii (Photographer: S.E. Rakotoarisoa)

Synonyms

Guillauminia rauhii (Reynolds) P.V.Heath

Etymology

rauhii: For Prof. Werner Rauh (1913–2000), German botanist in Heidelberg and specialist on Madagascan succulents.

Aloe rebmannii Lavranos

Etymology

rebmannii: For Prof. Norbert Rebmann (1948–), French university lecturer and aloe enthusiast.

Aloe reitzii Reynolds var. *reitzii*

Etymology

reitzii: For Mr F.W. Reitz, who discovered the plant and drew the author's attention to it.

Common names

bergaalwyn [9, 55] [Afrikaans]
Reitz's aloe [55] [English]

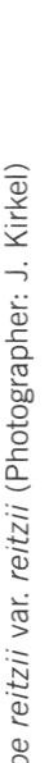

Aloe reitzii var. *reitzii* (Photographer: J. Kirkel)

Aloe reitzii Reynolds var. *vernalis* D.S.Hardy

Aloe reitzii var. *vernalis* (Photographer: SANBI, P. Joffe)

Etymology

reitzii: For Mr F.W. Reitz, who discovered the plant and drew the author's attention to it.

vernalis: For the flowering season in the spring months of August–September, from Latin 'vernalis' (pertaining to springtime).

Aloe rendilliorum L.E.Newton

Etymology

rendilliorum: For the Rendille tribe, in Kenya.

Aloe retrospiciens Reynolds

Aloe retrospiciens (Photographer: SANBI, J.J. Lavranos)

Synonyms

A. ruspoliana Baker var. *draceniformis* A.Berger

Etymology

draceniformis: For having the shape of a *Dracaena*, from Latin '-formis' (shaped).
retrospiciens: Looking back, for the orientation of buds and flowers, from Latin.
ruspoliana: For Prince Edugenio Ruspoli (1866–1893), Italian explorer and plant collector in northeast Africa, who collected the type.

Common names

daar burruk [55] [Unspecified language]
daar burruq [55] [Unspecified language]
daar burug [40] [Somali]

Aloe reynoldsii Letty

Aloe reynoldsii (Photographer: E.J. van Jaarsveld)

Etymology

reynoldsii: For Dr Gilbert W. Reynolds (1895–1967), Australian optometrist who emigrated to South Africa in 1902, who contributed significantly to the knowledge of *Aloe*.

Common names

yellow spineless aloe [95] [English]

Aloe rhodesiana Rendle

Aloe rhodesiana (Photographer: M.J. Kimberley)

Synonyms

A. eylesii Christian

Etymology

eylesii: For Frederick S. Eyles (1864–1937), English botanist and plant collector, who collected the type specimen in Zimbabwe in 1935.
rhodesiana: For the occurrence in southern Rhodesia (Zimbabwe).

Common names

Rhodesian aloe [7, 55, 126] [English]

Aloe ×*riccobonii* Borzí
(*A. arborescens* × *A. capitata*)

Etymology

riccobonii: For Vicenzo Riccobono (1861–1943), chief gardener of the Botanical Garden of Palermo, in Italy.

Aloe richardsiae Reynolds

Etymology

richardsiae: For Mrs H. Mary Richards (1885–1977), British plant collector, resident in East Africa 1952–1974.

Aloe richaudii Rebmann

Etymology

richaudii: For Philippe Richaudii, nurseryman in France, in whose greenhouse the plant grew.

Aloe rigens Reynolds & P.R.O.Bally var. *mortimeri* Lavranos

Etymology

mortimeri: For Prof. Keith V. Mortimer, British dentist and grower of succulents.
rigens: For the stiff leaves, from Latin 'rigens' (rigid).

Aloe rigens Reynolds & P.R.O.Bally var. *rigens*

Aloe rigens var. *rigens* (Photographer: SANBI, L.C. Leach)

Etymology

rigens: For the stiff leaves, from Latin 'rigens' (rigid).

Common names

daar merodi [55, 102] [Unspecified language]
dacar maroodi [55] [Unspecified language]

Aloe rivae Baker

Aloe rivae (Photographer: SANBI, G.W. Reynolds)

Etymology

rivae: For Dr Domenico Riva (c.1856–1895), Italian botanist, collector in northeast Africa, who collected the type.

Common names

argeesaa [40] [Oromo, West Central]

Aloe rivierei Lavranos & L.E.Newton

Aloe rivierei (Photographer: G. Orlando)

Synonyms

A. parvicoma Lavranos & Collen.

Etymology

parvicoma: For the few-leaved rosettes at the stem tips, from Latin 'parvus' (small), 'coma' (hair tuft, mane).
rivierei: For Fernando Riviere de Caralt (1904–1992), Spanish industrialist, grower of succulents and owner of the private botanical garden 'Pinya de Rosa'.

Aloe rodolphei J.-B.Castillon

Etymology

rodolphei: For Rodolphe Castillon, a grower of Madagascan succulent plants.

Aloe roeoeslii Lavranos & T.A.McCoy

Aloe roeoeslii (Photographer: G. Orlando)

Etymology

roeoeslii: For Walter Röösli, who collected the type with R. Hoffmann.

Aloe rosea (H.Perrier) L.E.Newton & G.D.Rowley

Synonyms

Lomatophyllum roseum H.Perrier

Etymology

rosea/roseum: For the rose-pink flowers, from Latin 'roseus' (rose-like).

Aloe rubrodonta T.A.McCoy & Lavranos

Etymology

rubrodonta: For the red teeth on the leaf margins, from Latin 'rubrus' (red) and Greek 'odous, odontos' (tooth).

Aloe rivierei

Aloe rubroviolacea Schweinf.

Aloe rubroviolacea (Photographer: SANBI, J.J. Lavranos)

Etymology

rubroviolacea: For the colour of the dry leaves, from Latin 'rubrus' (red), 'violaceus' (violet).

Aloe rugosifolia M.G.Gilbert & Sebsebe

Synonyms

A. otallensis Baker var. *elongata* A.Berger

Etymology

elongata: For the elongated inflorescence, from Latin 'elongatus' (elongate).
otallensis: For the occurrence at Otallo, in Ethiopia.
rugosifolia: For the rugose leaf surface, from Latin 'rugosus' (rugose), '-folius' (leaved).

Aloe rugosquamosa (H.Perrier) J.-B.Castillon & J.-P.Castillon

Synonyms

A. compressa H.Perrier var. *rugosquamosa* H.Perrier

Etymology

compressa: For the distichous (laterally compressed) leaf arrangements, from Latin 'compressus' (compressed).
rugosquamosa: For the rough upper leaf surface, from Latin 'ruga' (wrinkle), '-squamosus' (scaly).

Aloe ×*runcinata* A.Berger (*A. maculata* × *A. ferox*)

Synonyms

A. obscura A.Berger ex Schönland (nom. illegit.)

Etymology

obscura: Unresolved application, from Latin 'obscurus' (indistinct, obscure).
runcinata: For the serrate leaves, from Latin 'runcinatus' (runcinate).

Aloe rupestris Baker

Synonyms

A. nitens Baker (nom. illegit.)
A. pycnantha MacOwen

Etymology

nitens: Probably for the leaves, from Latin 'nitidus' (shining).
pycnantha: For the spines, from Greek 'pyknos' (dense), 'akantha' (thorn, spine).
rupestris: For the habitat, associated with rocks or cliffs, from Latin 'rupestris' (of rocks).

Aloe rupestris (Photographer: G. Nichols)

Common names

borselaalwyn [94, 99, 124] [Afrikaans]
bottle-brush aloe [55, 109] [English]
bottlebrush aloe [64, 94, 99, 124] [English]
inhlaba [53, 55, 64, 99] [Swati]
inhlabanzhlazi [49] [Zulu]
inkalane [50, 94, 99, 123] [Zulu]
inkhalane [53, 55, 123] [Zulu]
kraalaalwee [108] [Afrikaans]
kraalaalwyn [55, 108, 109, 123] [Afrikaans]
nkalane [50] [Zulu]
rock aloe [55, 99] [English]
uluphondonde [49] [Zulu]
umhlabandlanzi [94, 99] [Zulu]
umhlabandlazi [53, 55, 99] [Zulu]
umhlabanhazi [123] [Zulu]
umhlabanhlazi [49, 99] [Zulu]
umpondonde [51, 55] [Zulu]
undlampofu [49, 52] [Zulu]
uphondonde [49, 53, 55, 99, 123] [Zulu]
upondonde [94] [Zulu]

Aloe rupicola Reynolds

Etymology

rupicola: For the occurrence among rocks, from Latin 'rupes' (rocks, cliffs), '-cola' (inhabiting).

Aloe ruspoliana Baker

Synonyms

A. jex-blakeae Christian
A. stefaninii Chiov.

Aloe ruspoliana (Photographer: S. Demissew)

Etymology

jex-blakeae: For Lady Muriel Jex-Blake, who discovered the plant in 1936 in Kenya.
ruspoliana: For Prince Eugenio Ruspoli (1866–1893), Italian explorer and plant collector in northeast Africa, who collected the type.
stefaninii: For Giuseppe Stefanini (1882–1938), Italian naturalist, traveller and collector in eastern Africa, Ethiopia and Somalia.

Common names

dacar [55] [Unspecified language]

Aloe ruvuensis T.A.McCoy & Lavranos

Etymology

ruvuensis: For the occurrence near the Ruvu River, in Tanzania.

Aloe sabaea Schweinf.

Synonyms

A. gillilandii Reynolds

Etymology

gillilandii: For Prof. H.B. Gilliland (fl. 1952) from the University of Malay, Singapore, who first discovered the plants, in Arabia.
sabaea: Probably commemorating the state of Saba (Sheba).

Aloe sabaea (Photographer: SANBI)

Aloe sakarahensis Lavranos & M.Teissier subsp. *pallida* (Rauh) Lavranos & M.Teissier

Synonyms

A. prostrata (H.Perrier) L.E.Newton & G.D.Rowley subsp. *pallida* Rauh

Etymology

pallida: For the pale flower, from Latin 'pallidus' (pale).
prostrata: For the spreading leaves, from Latin 'prostratus' (prostrate, creeping).
sakarahensis: For the occurrence in the Sakahara Forest, in Madagascar.

Aloe sakarahensis Lavranos & M.Teissier subsp. *sakarahensis*

Etymology

sakarahensis: For the occurrence in the Sakahara Forest, in Madagascar.

Aloe saudiarabica T.A.McCoy

Etymology

saudiarabica: For the occurrence in Saudi Arabia.

Aloe saundersiae (Reynolds) Reynolds

Aloe saundersiae (Photographer: G. Nichols)

Synonyms

A. minima J.M.Wood (nom. illegit.)
Leptaloe saundersiae Reynolds

Etymology

minima: For the small size of the plant, from Latin 'minimus' (very small, smallest).

saundersiae: For Lady Katherine Saunders (née Wheelright) (1824–1901), English plant collector and botanical artist in South Africa, mother of Charles James Renault Saunders (1857–1935), explorer and collector in Rhodesia (Zimbabwe) and Mozambique. According to Reynolds[101] it was collected by Lady Saunders in the 1930s in KwaZulu-Natal, South Africa, but since she died in 1901 it was probably collected by her son.

Aloe scabrifolia L.E.Newton & Lavranos

Etymology

scabrifolia: For the rough leaves, from Latin 'scabrum' (rough), '-folius' (leaved).

Aloe schelpei Reynolds

Aloe schelpei (Photographer: SANBI, L.C. Leach)

Etymology

schelpei: For Prof. Edmund A.C.L.E. Schelpe (1924–1985), South African botanist at the University of Cape Town, who collected the type.

Aloe schilliana L.E.Newton & G.D.Rowley

Synonyms

Lompatophyllum viviparum H.Perrier

Etymology

schilliana: For Prof. Rainer Schill, German botanist at Heidelberg University.

viviparum: Unresolved application, from Latin 'viviparus' (viviparous).

Aloe ×*schimperi* Tod.
(*A. maculata* × *A. striata*)

Aloe ×*schimperi* (Photographer: E.J. van Jaarsveld)

Synonyms

A. ×*paxii* A.Terracc.

Etymology

paxii: Unresolved application, possibly for the botanist Ferdinand A. Pax (1858–1942).

schimperi: For Georg W. Schimper (1804–1878), German botanist and plant collector, who lived and became nationalised in Abyssinia.

Common names

makbontaalwyn [109] [Afrikaans]
spotted aloe hybrid [109] [English]

Aloe schoelleri Schweinf.

Aloe schoelleri (Photographer: G. Orlando)

Etymology

schoelleri: For Max Shoeller, German ethnologist who travelled widely in Africa.

Aloe ×*schoenlandii* Baker
(*A. striata* × *A. maculata*)

Etymology

schoenlandii: For Dr Selmar Schönland (1860–1940), German-born botanist who emigrated to South Africa in 1899 and became director of the Albany Museum in Grahamstown.

Aloe schomeri Rauh

Etymology

schomeri: For Menko Schomerus, a mine-owner in Madagascar.

Aloe schweinfurthii Baker

Synonyms

A. barteri Baker, p.p. (leaf only)
A. barteri Baker var. *lutea* A.Chev.

Etymology

barteri: For Charles Barter (fl. 1857–1859) British gardener, foreman of the Regent's Park gardens of the Royal Botanic Society, London who joined the second Niger Expedition of W. Baikie and collected the type.

Aloe schweinfurthii (Photographer: SANBI, G.W. Reynolds)

lutea: For the yellow flowers, from Latin 'luteus' (yellow).
schweinfurthii: For Dr Georg Schweinfurth (1836–1925), German botanist, geographer and explorer of northeast Africa and Arabia.

Common names

ranga [55, 102] [Zande]
rangambala [55] [Zande]
rangambia [102] [Zande]

Aloe scobinifolia Reynolds & P.R.O.Bally

Etymology

scobinifolia: For the rough leaves, from Latin 'scobina' (rasp), '-folius' (leaved).

Common names

daar [55, 102] [Somali]

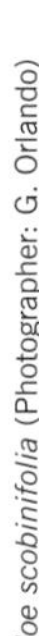

Aloe scobinifolia (Photographer: G. Orlando)

Aloe scorpioides L.C.Leach

Aloe scorpioides (Photographer: SANBI, L.C. Leach)

Etymology

scorpioides: For the shape of the peduncle, from Latin 'scorpioides' (scorpioid).

Aloe secundiflora Engl. var. *secundiflora*

Aloe secundiflora var. *secundiflora* (Photographer: SANBI)

Synonyms

A. engleri A.Berger
A. floramaculata Christian
A. marsabitensis I.Verd. & Christian

Etymology

engleri: For Prof. Dr Heinrich Gustav Adolf Engler (1844–1930), German botanist and plant collector, who collected it in 1902.
floramaculata: For the spotted flowers, from Latin 'flos' (flower), 'maculatus' (spotted).
marsabitensis: For the occurrence at Marsabit, in Kenya.
secundiflora: For the secund flowers (directed to one side), from Latin 'secundus' (secund), '-florus' (flowered).

Common names

achuka [102] [Karamojong]
echichuviwa [55] [Turkana]

echuchuka [55] [Turkana]
echuchukua [102] [Karamojong]
endadaijoko [55, 84] [Maasai]
endadaiyoku [55] [Unspecified language]
esugoroi [55] [Maasai]
golonje [55, 84] [Unspecified language]
harguessa [55] [Borana]
joloji [55] [Chidigo]
kigaka [55] [Lulogooli]
kiluma [55, 84] [Kamba]
kirumi [55] [Gikuyu]
kisimamleo [55] [Swahili]
kisimando [55] [Swahili]
kitori [55] [Kigiryama]
lineke [55] [Lulogooli]
mũgwanũgũ [55] [Gikuyu]
mukumi [55] [Kiembu]
ogara [55] [Dholuo]
olkos [55] [Pökoot]
osuguroi [55] [Maasai]
sikorowet [55] [Pökoot]
suk' ro-i [55, 84] [Samburu]
sukoroi [55] [Samburu]
tangaratuet [55] [Nandi]
tangaratwet [55] [Kipsigis]
thugurui [55] [Gikuyu]
tolkos [55] [Pökoot]

Aloe secundiflora Engl. var. *tweediae* (Christian) Wabuyele

Synonyms

A. tweediae Christian

Etymology

secundiflora: For the secund flowers (directed to one side), from Latin 'secundus' (secund), '-florus' (flowered).
tweediae: For E. Marjorie Tweedie, British artist and collector resident in Kenya from 1918 onwards, who discovered and collected the plants in Uganda.

Common names

achuka [102] [Pökoot]
chokokwet [102] [Kupsabiny]
echuchukua [55, 102] [Karamojong]
etchuka [55, 102] [Pökoot]

Aloe seretii De Wild.

Aloe seretii (Photographer: SANBI, L.C .Leach)

Etymology

seretii: For Felix Seret (fl. 1905–1909), Belgian forestry officer and plant collector in the now Democratic Republic of Congo.

Aloe serriyensis Lavranos

Etymology

serriyensis: For the occurrence at the village of Serriya, in Yemen.

Aloe shadensis Lavranos & Collen.

Etymology

shadensis: For the occurrence in Jabal Shada, in Saudi Arabia.

Aloe sheilae Lavranos

Synonyms

A. cephalophora Lavranos & Collen.

Etymology

cephalophora: For the capitate inflorescences, from Greek 'kephale' (head), '-phoros' (carrying).

sheilae: For Iris Sheila Collenette (1927–), English amateur botanist, well-known collector and researcher of Arabian succulents.

Aloe silicicola H.Perrier

Etymology

silicicola: For the preferred habitat, from Latin 'silicis' (silica), '-cola' (inhabiting).

Aloe simii Pole-Evans

Etymology

simii: For Dr T.R. Sim (1858–1938), horticulturalist and botanist in South Africa, who first collected the plant.

Aloe sinana Reynolds

Aloe sinana (Photographer: SANBI, G.W. Reynolds)

Etymology

sinana: For the occurrence at Debre Sina, in Ethiopia.

Aloe sinkatana Reynolds

Aloe sinkatana (Photographer: SANBI, G.W. Reynolds)

Etymology

sinkatana: For the occurrence at Sinkat, in Sudan.

Common names

kalandoy [55, 102] [Bedawi]

Aloe sladeniana Pole-Evans

Synonyms

A. carowii Reynolds

Etymology

carowii: For Mr R. Carow, who discovered it at Nauchas, in Namibia.

sladeniana: For William Percy Sladen (?–1900), British naturalist and financial benefactor of the expedition on which this plant was discovered.

Aloe sobolifera (S.Carter) Wabuyele

Synonyms

A. leachii Reynolds
A. secundiflora Engl. var. *sobolifera* S.Carter

Etymology

leachii: For Leslie (Larry) C. Leach (1909–1996), English-born engineer and botanist in Zimbabwe, later in South Africa.
secundiflora: For the secund flowers (directed to one side), from Latin 'secundus' (secund), '-florus' (flowered).
sobolifera: For the offsetting nature of the plants, from Latin 'soboles' (branch, offspring), '-fer' (carrying).

Aloe socialis (H.Perrier) L.E.Newton & G.D.Rowley

Synonyms

Lomatophyllum sociale H.Perrier

Etymology

sociale/socialis: For the clustering habit, from Latin 'socialis' (social).

Aloe somaliensis W.Watson var. *marmorata* Reynolds & P.R.O.Bally

Etymology

marmorata: For the pattern of leaf markings, giving a marbled effect, from Latin 'marmoratus' (marbled).
somaliensis: For the occurrence in Somalia.

Aloe somaliensis W.Watson var. *somaliensis*

Etymology

somaliensis: For the occurrence in Somalia.

Common names

daar biyu [55, 102] [Unspecified language]
dacar biyu [55] [Unspecified language]

Aloe soutpansbergensis I.Verd.

Aloe soutpansbergensis (Photographer: SANBI, D.S. Hardy)

Etymology

soutpansbergensis: For the occurrence on the Soutpansberg in Limpopo Province, South Africa.

Aloe speciosa Baker

Aloe speciosa (Photographer: G.F. Smith)

Etymology

speciosa: For the showy, beautiful inflorescences, from Latin 'speciosus' (beautiful).

Common names

beautiful aloe [99] [English]
ikhala [94, 99] [Xhosa]
slapare-aalwyn [55] [Afrikaans]
slaphoringaalwyn [53, 55, 62, 94, 99, 109, 123, 124] [Afrikaans]
slapoor [53, 99] [Afrikaans]
slapooraalwyn [55, 124] [Afrikaans]
spaansaalwee [108] [Afrikaans]
spaansaalwyn [55, 62, 108, 109, 123, 124] [Afrikaans]
spanareaalwyn [53, 55, 62, 99] [Afrikaans]
spanarei-aalwee [108] [Afrikaans]
spanerei-aalwyn [55, 108] [Afrikaans]
spansaalwyn [53, 55, 99] [Afrikaans]
tilt-head aloe [94, 99, 109, 123, 124] [English]
wildeaalwyn [55, 62, 108] [Afrikaans]

Aloe spectabilis Reynolds

Aloe spectabilis (Photographer: G.F. Smith)

Etymology

spectabilis: For the general appearance, from Latin 'spectabilis' (showy).

Common names

ikala [50] [Zulu]
ilikala [50] [Zulu]
inhlaba [94, 99] [Zulu]
mokqala [50] [Zulu]
Natal aloe [94, 99, 124] [English]
Natalaawyn [94, 99, 124] [Afrikaans]
umhlaba [49, 99, 101] [Zulu]
umkala [50] [Zulu]

Products

Natal aloes: Prepared from the leaf exudate[92].

Aloe spicata L.f.

Aloe spicata (Photographer: G. Nichols)

Synonyms

A. sessiliflora Pole-Evans

Etymology

sessiliflora: For the sessile flowers, from Latin 'sessilis' (sessile), '-florus' (flowered).
spicata: For the long and densely-flowered spike-like inflorescences, from Latin 'spicatus' (spicate).

Common names

aloé dos libombos [4] [Portuguese]
bottle brush aloe [126] [English]
Gazaland aloe [55] [English]
inhlaba [95, 64] [Swati] [95] [Zulu]
Lebombo aloe [4, 55, 57, 64, 86, 95, 109, 123, 124] [English]
Lebombo-aalwyn [55, 86, 95, 109, 123, 124] [Afrikaans]
Lebomboaalwyn [57] [Afrikaans]
spaansaalwee [108] [Afrikaans]
spaansaalwyn [108] [Afrikaans]
spanarei-aalwee [108] [Afrikaans]
spanarei-aalwyn [108] [Afrikaans]
tshikhopha [57] [Venda]
umhlaba [49] [Zulu]
wildeaalwyn [108] [Afrikaans]

Aloe ×spinosissima Jahand (*A. humilis* var. *echinata* × *A. arborescens*)

Etymology

spinosissima: For being very spiny, from Latin 'spinosus' (spiny).

Aloe splendens Lavranos

Synonyms

A. doei Lavranos
A. doei Lavranos var. *lavranosii* Marn.-Lap.

Etymology

doei: For Brian Doe, director of the Department of Antiquities in Aden, Yemen, who found the plant during an excursion with the author.
lavranosii: For John J. Lavranos (1926–) Greek insurance broker, botanist and collector of succulents throughout southern and eastern Africa, Arabia and Madagascar.
splendens: For the flower colour, from Latin 'splendens' (brilliant).

Aloe ×spuria A.Berger

Etymology

spuria: Because it could be a hybrid, from Latin 'spurius' (doubtful, false).

Aloe squarrosa Baker ex Balf.f.

Aloe squarrosa (Photographer: G. Orlando)

Synonyms

A. concinna Baker (nom. illegit.)
A. zanzibarica Milne-Redh.

Etymology

concinna: For the appearance, from Latin 'concinnus' (neat, pretty, elegant).
squarrosa: For the rough leaf surface, from Latin 'squarrosus' (spreading, recurved).
zanzibarica: For the occurrence in Zanzibar, Tanzania.

Aloe steudneri Schweinf.

Aloe steudneri (Photographer: G. Orlando)

Etymology

steudneri: For Dr H. Steudner (1832–1863), botanist and explorer in north-east Africa who collected the type.

Aloe striata Haw.

Aloe striata (Photographer: J.E. Burrows)

Synonyms

A. albocincta Haw.
A. hanburiana Naudin
A. paniculata Jacq.
A. rhodocincta Baker
A. striata Haw. var. *oligospila* Baker
A. striata Haw. var. *rhodocincta* (Baker) Trel.

Etymology

albocincta: For the white-edged leaves, from Latin 'albus' (white), 'cinctus' (encircled).
hanburiana: For Sir Thomas Hanbury (1832–1907) who founded the Hanbury Botanic Gardens (La Mortola) near Ventimiglia in Italy, in 1867.

Aloe squarrosa

oligospila: For being few-haired, from Greek 'oligos' (few) and Latin 'pilus' (hair).
paniculata: For the inflorescence, from Latin 'paniculatus' (paniculate).
rhodocincta: For the pale reddish leaf margin, from Greek 'rhodos' (rose-red) and Latin 'cinctus' (encircled).
striata: For the lines on the leaves, from Latin 'striatus' (striate).

Common names

blou-aalwee [108] [Afrikaans]
blouaalwyn [55, 62, 109, 123] [Afrikaans]
blou-aalwyn [81, 108] [Afrikaans]
coral aloe [9, 55, 56, 62, 98, 106, 108, 109, 123] [English]
gladdeblaaraalwyn [9, 55, 62, 106, 123] [Afrikaans]
icena [47] [Zulu]
ilicena [47] [Zulu]
ingcelwane [55] [Xhosa]
koraalaalwyn [55, 81, 109] [Afrikaans]
makaalwyn [55, 62, 108, 123] [Afrikaans]
streepaalwyn [55, 62, 108, 109, 123] [Afrikaans]
Uitenhaagse aalwyn [9, 55] [Afrikaans]
vaalblaaraalwyn [55, 62, 108, 123] [Afrikaans]
vaalblaar-aalwyn [98] [Afrikaans]

Aloe striatula Haw. var. *caesia* Reynolds

Aloe striatula var. *caesia* (Photographer: N.R. Crouch)

Etymology

caesia: For the blue-grey colour of the leaves, perhaps erroneously as in the diagnosis the author states 'foliis caesio-viridulis', but the explanation reads 'milky-green leaves', from Latin 'caesius' (light blue).
striatula: For the thin green parallel lines on the leaf sheaths, from Latin 'striatus' (striate).

Common names

lekhala [58] [Sotho, Southern]
mohalakane [58] [Sotho, Southern]

Aloe striatula Haw. var. *striatula*

Aloe striatula var. *striatula* (Photographer: N.R. Crouch)

Synonyms

A. aurantiaca Baker
A. cascadensis Kuntze
A. macowanii Baker
A. subinermis Lem.

Etymology

aurantiaca: For the colour of the flower, from Latin 'aurantiacus' (orange).
cascadensis: For occurring at a small waterfall near East London, in South Africa.

macowanii: For Prof. Peter MacOwan (1830–1909), botanist and curator of the Cape Government Herbarium and Cape Town Botanic Gardens, in South Africa.
striatula: For the thin green parallel lines on the leaf sheaths, from Latin 'striatus' (striate).
subinermis: For being almost unarmed, from Latin 'sub' (almost), 'inermis' (unarmed).

Common names

heiningaalwyn [9, 55] [Afrikaans]
mohalakane [96] [Sotho, Southern]
seholobe [96] [Sotho, Southern]
stripe-sheathed narrow-leaved aloe [55] [English]
unqcelwane [55] [Xhosa]

Aloe suarezensis H.Perrier

Aloe suarezensis (Photographer: SANBI, G.W. Reynolds)

Etymology

suarezensis: For the occurrence in the region of Diego Suarez (Antsiranana), in Madagascar.

Aloe succotrina Weston

Aloe succotrina (Photographer: A.W. Klopper)

Synonyms

A. perfoliata L. var. *purpurascens* Aiton
A. perfoliata L. var. *succotrina* (Lam.) Aiton
A. perfoliata L. var. ξ L.
A. purpurascens (Aiton) Haw.
A. sinuata Thunb.
A. soccotorina Schult. & Schult.f.
A. soccotrina Garsault
A. socotrina DC.
A. socotrina DC. var. [β] *purpurascens* (Aiton) Ker Gawl.
A. succotrina Lam.
A. succotrina Lam. var. *saxigena* A.Berger
A. vera Mill. (nom. illegit.)

Etymology

perfoliata: For the stem passing through the leaves, i.e. the leaves are amplexicaul, from Latin 'per' (through), 'folia' (leaf).
purpurascens: For the leaves turning purple when dry, from Latin 'purpurascens' (becoming purple).

Aloe striatula var. *striatula*

saxigena: For occurring between stones, from Latin 'saxum' (rock), '-genus' (born).

sinuata: Probably for the leaves, from Latin 'sinuatus' (waved, sinuate).

soccotorina/soccotrina/socotrina/succotrina: Either for the plant being thought to be the source of the drug socotrine aloes and originating from Socotra (although it grows wild only in the extreme south-western Cape, in South Africa) or for the compound word meaning 'succus' (sap), 'citrinus' (lemon-yellow). Although the purple juice is characteristic of the plant[53, 101] it was reported that it turns yellow when it dries[67].

vera: The true aloe, from Latin 'vera' (in truth, real).

Common names

açevar [30] [Spanish]
áloe estriada [73] [Spanish]
áloe socotrina [73] [Spanish]
asevar [30] [Spanish]
bergaalwee [108] [Afrikaans]
bergaalwyn [55, 62, 108, 109] [Afrikaans]
bole-siyah [55] [Persian group]
Bombay aloe [66] [English]
chennanayakam [55] [Malayalam]
eliya [55] [Deccan]
elva [55] [Deccan]
hierba del acíbar [73] [Spanish]
ilva [55] [Hindi]
kalu-bolam [55] [Sindhi]
karibolam [55] [Sindhi]
kariya-polam [55] [Tamil]
mocha aloe [66] [English]
moshabbar [55] [Bengali]
musabbar [55] [Hindi]
musanbar [55] [Deccan]
mushambaram [55] [Telugu]
sabila [55] [English]
sabir [55] [Arabic]
sibr [55] [Arabic]
Socotrine aloes [55] [English]
Table Mountain aloe [55, 109] [English]
Tafelbergaalwyn [55, 109] [Afrikaans]
Turkey aloe [66] [English]
yalva [55] [Hindi]
yeliyo [55] [Gujarati]
yerba babose [30] [Spanish]
yerba del acibar [30] [Spanish]
Zanzibar aloe [66] [English]

Aloe suffulta Reynolds

Aloe suffulta (Photographer: N.R. Crouch)

Etymology

suffulta: Reference to the weak and slender inflorescences which are always supported by the surrounding vegetation, from Latin 'suffultus' (supported).

Common names

climbing-flower aloe [55, 126] [English]

Aloe suprafoliata Pole-Evans

Etymology

suprafoliata: Refers to the leaves of the young plants that are seemingly situated on top of each other in two rows, resembling the pages of an open book, from Latin 'supra' (above), 'foliatus' (leaved).

Aloe suprafoliata (Photographer: G. Nichols)

Common names

boekaalwyn [55, 62, 95, 109, 123] [Afrikaans]
book aloe [55, 95, 109] [English]
icena [47, 95] [Zulu]
ilicena [47] [Zulu]
inhlaba [95] [Swati]
inkalane [50] [Zulu]
umhlabanhlazi [49, 95] [Zulu]

Aloe suzannae Decary

Etymology

suzannae: For Suzanne Decary, daughter of the author.

Common names

vahondrano [90] [Unspecified language]

Aloe suzannae (Photographer: G. Orlando)

Aloe swynnertonii Rendle

Synonyms

A. chimanimaniensis Christian
A. melsetterensis Christian

Etymology

chimanimaniensis: For the occurrence in the Chimanimani Mountains, eastern Zimbabwe.
melsetterensis: For the occurrence near Melsetter, in Zimbabwe.
swynnertonii: For Charles F.M. Swynnerton (1877–1939), English zoologist and naturalist, botanical explorer in Mozambique, Tanzania and Zimbabwe, who collected the first specimens in Zimbabwe.

Common names

chitembwe [55] [Nyanja]
lichongwe [55] [Nyanja]

Aloe swynnertonii (Photographer: M.J. Kimberley)

Aloe tauri (Photographer: M.J. Kimberley)

nasi [55] [Nyanja]
Swynnerton's aloe [55] [English]
Swynnerton's spotted leaf aloe [55, 126] [English]

Aloe tartarensis T.A.McCoy & Lavranos

Etymology

tartarensis: For the occurrence at Tartar Falls, in Kenya.

Aloe tauri L.C.Leach

Etymology

tauri: For E.J. Bullock of Bulawayo, Rhodesia (Zimbabwe), a student of the genus, who discovered the species, from Latin/Greek 'taurus' (bull).

Common names

Bullock's bottle brush aloe [126] [English]

Aloe teissieri Lavranos

Synonyms

A. andohahelensis J.-B.Castillon

Etymology

andohahelensis: For the occurrence on the Massif of Andohahela, in Madagascar.
teissieri: For Marc Teissier, French horticulturalist and curator of the private botanical garden 'Les Cèdres' at Saint Jean Cap Ferrat near Nice, in France.

Common names

vaho [29] [Malagasy]

Aloe tenuifolia Lam.

Etymology

tenuifolia: For the thin leaves, from Latin 'tenuis' (slender, thin), 'folius' (leaved).

Aloe tenuior Haw.

Aloe tenuior (Photographer: J.E. Burrows)

Synonyms

A. tenuior Haw. var. *decidua* Reynolds
A. tenuior Haw. var. *densiflora* Reynolds
A. tenuior Haw. var. *glaucescens* Zahlbr.
A. tenuior Haw. var. *rubriflora* Reynolds
A. tenuior Haw. var. *viridifolia* van Jaarsv.

Etymology

decidua: For the deciduous leaves, from Latin 'deciduus' (deciduous).
densiflora: For the dense inflorescences, from Latin 'densus' (dense), '-florus' (flowered).
glaucescens: For the blue-green colour, from Latin 'glaucus' (glaucous), '-escens' (becoming).
rubriflora: For the red flowers, from Latin 'rubrus' (red), '-florus' (flowered).
tenuior: Referring to the slender branches, from Latin 'tenuis' (slender).
viridifolia: For the green leaves, from Latin 'viridis' (green), '-folia' (leaved).

Common names

empofu [95] [Xhosa] [95] [Zulu]
fence aloe [55] [English]
green-sheathed narrow-leaved aloe [53, 55] [English]
heuningaalwyn [95] [Afrikaans]
ikalana [107] [Unspecified language]
ikhalana [76, 99] [Xhosa]
ikhalene [53, 55, 95, 101] [Zulu]
inhlaba [95] [Xhosa] [95] [Zulu]
inhlaba empofu [55] [Zulu]
inkalane [50] [Zulu]
intelezi [53, 55, 95, 101] [Xhosa]
umjinqa [55] [Xhosa]

Aloe tewoldei M.G.Gilbert & Sebsebe

Etymology

tewoldei: For Tewolde-Bergan Gebre-Egziabher, Ethiopian botanist and one of the joint leaders of the Ethiopian Flora Project.

Aloe thompsoniae Groenew.

Aloe thompsoniae (Photographer: SANBI, G.W. Reynolds)

Etymology

thompsoniae: For Dr Sheila Thompson (née Clifford) (fl. 1930s) of Haenertsberg, South Africa, who collected it in 1924.

Common names

kleingrasaalwyn [55] [Afrikaans]
Thompson's aloe [55] [English]

Aloe thorncroftii Pole-Evans

Aloe thorncroftii (Photographer: SANBI, G.W. Reynolds)

Etymology

thorncroftii: For George Thorncroft (1874–1934), keen gardener and collector in Barberton, South Africa, who first collected it.

Aloe thraskii Baker

Aloe thraskii (Photographer: G. Nichols)

Synonyms

A. candelabrum Engl. & Drude (nom. illegit.)

Etymology

candelabrum: For the appearance of the inflorescence, which resembles a candlestick or candelabrum, from Latin.
thraskii: For a Mr Thrask.

Common names

coast aloe [55] [English]
dune aloe [53, 55, 94, 99, 123, 124] [English]
ikhala [94, 99] [Xhosa]
imihlaba [55] [Zulu]
inhlaba [55] [Zulu]
strand aloe [53, 55, 94, 99, 101, 123, 124] [English]
strandaalwyn [9, 53, 55, 62, 94, 99, 123] [Afrikaans]
strandveldaalwyn [124] [Afrikaans]
umhlaba [49, 53, 55, 94, 99] [Zulu]

Aloe tomentosa Deflers

Etymology

tomentosa: For the hairy flowers, from Latin 'tomentosus' (felted, covered in matted hairs).

Aloe tormentorii (Marais) L.E.Newton & G.D.Rowley

Synonyms

Lomotaophyllum tormentorii Marais

Etymology

tormentorii: For the type locality, Gunner's Quoin on Round Island, Mauritius, from Latin 'tormentum' (a military engine for discharging missiles).

Aloe tororoana Reynolds

Etymology

tororoana: For the occurrence on Tororo Rock, in Uganda.

Aloe thorncroftii

Aloe torrei I.Verd. & Christian

Aloe torrei (Photographer: SANBI, L.C. Leach)

Etymology

torrei: For António Rocha da Torre (1904–1995), Portuguese biologist and pharmacist.

Aloe trachyticola (H.Perrier) Reynolds var. *multifolia* J.-B.Castillon

Etymology

multifolia: For the numerous leaves, from Latin 'multi-' (many), '-folia' (leaved).
trachyticola: For the habitat, from English/ French 'trachyte' (trachyte rock) and Latin '-cola' (inhabiting).

Aloe trachyticola (H.Perrier) Reynolds var. *trachyticola*

Synonyms

A. capitata Baker var. *trachyticola* H.Perrier

Etymology

capitata: For the head-like inflorescence, from Latin 'capitatus' (capitate).

Aloe trachyticola var. *trachyticola* (Photographer: R.R. Klopper)

trachyticola: For the habitat, from English/ French 'trachyte' (trachyte rock) and Latin '-cola' (inhabiting).

Aloe transvaalensis Kuntze

Synonyms

A. laxissima Reynolds
A. transvaalensis Kuntze var. *stenacantha* Groenew.

Etymology

laxissima: For the laxly-flowered inflorescence, from Latin 'laxus' (lax), superlative.
stenacantha: For the narrow spines, from Greek 'stenos' (narrow), 'akanthos' (spine).
transvaalensis: For the occurrence in the former Transvaal Province (now split into Gauteng, Limpopo, North-West and Mpumalanga Provinces), in South Africa.

Aloe torrei

Aloe transvaalensis (Photographer: R.R. Klopper)

Aloe trichosantha subsp. trichosantha (Photographer: SANBI, J.J. Lavranos)

Aloe trichosantha A.Berger subsp. *longiflora* M.G.Gilbert & Sebsebe

Etymology

longiflora: For the longer flowers, from Latin 'longus' (long), '-florus' (flowered).
trichosantha: For the hairy perianth, from Greek 'trichos' (hair), 'anthos' (flower).

Aloe trichosantha A.Berger subsp. *trichosantha*

Synonyms

A. percrassa Schweinf. var. *albo-picta* Schweinf.

Etymology

albo-picta: For the white spots on the leaves, from Latin 'albus' (white), 'pictus' (painted).
percrassa: For the succulent leaves, from Latin 'per-' (very), 'crassus' (thick).
trichosantha: For the hairy perianth, from Greek 'trichos' (hair), 'anthos' (flower).

Common names

argeesaa [40] [Oromo, West Central]
daar [40] [Somali]
daar lebi [40] [Somali]
daar merodi [40] [Somali]
erreh [55] [Unspecified language]
genenoo [40] [Hadiyya]
godole uta [40] [Wolaytta]
reetii [40] [Oromo, West Central]
wend `riet [40] [Amharic]

Aloe trigonantha L.C.Leach

Etymology

trigonantha: For the markedly trigonous perianth, from Greek 'trigonous' (triangular), 'anthos' (flower).

Aloe tulearensis T.A.McCoy & Lavranos

Etymology

tulearensis: For the occurrence near Tuléar (Toliara), in Madagascar.

Aloe turkanensis Christian

Etymology

turkanensis: For the occurrence in the Turkana District, Kenya.

Aloe ×*ucriae* A.Terracc. (*A. arborescens* × *A. pluridens*)

Synonyms

A. arborescens Mill. var. *ucriae* (A.Terracc.) A.Berger

Etymology

arborescens: For becoming tree-like, from Latin 'arbor' (tree), although the plant is not a tree but a large, much-branched shrub[53].

ucriae: For Bernardino da Ucria (1739–1796), Italian Franciscan monk and botanist, and curator of the Botanic Garden of Palermo.

Aloe ukambensis Reynolds

Etymology

ukambensis: For the occurrence in Ukambani District (now Kitui and Machakos Districts), in Kenya.

Aloe umfoloziensis Reynolds

Aloe umfoloziensis (Photographer: G. Nichols)

Etymology

umfoloziensis: For the occurrence near the Black and White Umfolozi Rivers, in South Africa.

Common names

icena [47] [Zulu]
ilicena [47] [Zulu]

Aloe vacillans Forssk.

Synonyms

A. audhalica Lavranos & D.S.Hardy
A. dhalensis Lavranos

Etymology

audhalica: For the occurrence on the Audhali Plateau, in Yemen.

dhalensis: For the occurrence at Dhala, in Yemen.
vacillans: Probably for the habit, as it becomes decumbent, from Latin 'vacillans' (swinging to and fro).

Aloe vallaris L.C.Leach

Aloe vallaris (Photographer: SANBI, L.C. Leach)

Etymology
vallaris: For the habitat, from Latin 'vallaris' (of walls).

Aloe vanbalenii Pillans

Aloe vanbalenii (Photographer: SANBI, L.C. Leach)

Etymology
vanbalenii: For Jan C. van Balen (1894–1956), horticulturalist and former director of Parks of Johannesburg, South Africa, who first collected it.

Common names
icena lamatshe [47] [Zulu]
icenalamatshe [55, 95] [Zulu]
icenandhlovu [53] [Zulu]
icenlandhlovu [55, 95] [Zulu]
ilicena lamatshe [47] [Zulu]
inhlahlwane [49, 95] [Zulu]
lihlala [95] [Swati]
Van Balen's aloe [95] [English]

Aloe vandermerwei Reynolds

Synonyms
A. angustifolia Groenew. (nom. illegit.)

Etymology
angustifolia: For the narrow leaves, from Latin 'angustus' (narrow), '-folius' (leaved).
vandermerwei: For Dr Frederick Z. van der Merwe (1894–1968), South African medical inspector of schools, and specialist in *Aloe* and *Scilla*.

Aloe vanrooyenii Gideon F.Sm. & N.R.Crouch

Aloe vanrooyenii (Photographer: G.F. Smith)

Etymology

vanrooyenii: For Mr Pieter van Rooyen of Greytown, South Africa, who prompted further investigation of wild populations of the species[111, 112].

Aloe vaombe Decorse & Poiss. var. *poissonii* Decary

Etymology

poissonii: For Dr Henri L. Poisson (1877–1963), French veterinary surgeon and botanist, resident in Madagascar 1916–1954.

vaombe: For the local common name of the plants in Madagascar.

Aloe vaombe Decorse & Poiss. var. *vaombe*

Aloe vaombe var. *vaombe* (Photographer: J.-P. Castillon)

Etymology

vaombe: For the local common name of the plants in Madagascar.

Common names

vahombe [55, 90, 91] [Unspecified language]

Aloe vaotsanda Decary

Aloe vaotsanda (Photographer: S.E. Rakotoarisoa)

Etymology

vaotsanda: For the local common name of the plants in Madagascar.

Common names

vahotsanda [91] [Unspecified language]

Aloe variegata L.

Aloe variegata (Photographer: J.C. Kruger)

Synonyms

A. ausana Dinter
A. punctata Haw.
A. variegata L. var. *haworthii* A.Berger

Etymology

ausana: For the occurrence at Aus, in Namibia.
haworthii: For Adrian H. Haworth (1768–1833), English zoologist and botanist, succulent plant specialist.
punctata: For the spotted leaves, from Latin 'punctus' (dot).
variegata: For the spotted leaves, from Latin 'variegatus' (variegated).

Common names

ba toba xha [99] [Unspecified language]
bontaalwyn [55, 60, 62, 71, 75, 81, 108] [Afrikaans]
bontbees [93] [Xiri]
choje [55] [Unspecified language]
degoree [55] [Unspecified language]
kanniedood [9, 55, 60, 62, 75, 78, 81, 98, 101, 106, 108, 123] [Afrikaans] [93] [Xiri]
kanniedood aloe [56] [English]
Luckhoffaalwyn [55, 62] [Afrikaans]
Luckhoffse aalwyn [9, 55] [Afrikaans]
partridge aloe [60, 75] [English]
partridge breast aloe [9, 55, 62] [English]
partridge-breast aloe [56, 101] [English]
tiger aloe [56] [English]
variegated aloe [71, 106] [English]

Aloe venenosa Engl.

Etymology

venenosa: For being poisonous, from Latin 'venenosus' (very poisonous).

Aloe vera (L.) Burm.f.

Aloe vera (Photographer: G.F. Smith)

Synonyms

A. barbadensis Mill.
A. barbadensis Mill. var. *chinensis* Haw.

A. chinensis Steud. ex Baker
A. elongata Murray
A. flava Pers.
A. indica Royle
A. lanzae Tod.
A. maculata Forrsk. (nom. illegit.)
A. perfoliata L. var. [γ] *barbadensis* (Mill.) Aiton
A. perfoliata L. var. [λ] *vera* Willd.
A. perfoliata L. var. [π] *vera* L.
A. rubescens DC.
A. sabila Karw. ex Steud.
A. variegata Forssk (nom. illegit.)
A. vera L. var. *chinensis* (Steud. ex Baker) Baker
A. vera L. var. *lanzae* (Tod.) Baker
A. vulgaris Lam.

Etymology

barbadensis: For the occurrence in Barbados, West Indies.
chinensis: For the occurrence in China.
elongata: Probably for the elongated inflorescence, from Latin 'elongatus' (elongate).
flava: For the colour of the flowers, from Latin 'flavus' (yellow).
indica: For the occurrence in India.
lanzae: For Dr Domenico Lanza, botanist at the Palermo botanical garden.
maculata: For the spotted leaves, from Latin 'maculatus' (spotted).
perfoliata: For the stem passing through the leaves, i.e. the leaves are amplexicaul, from Latin 'per' (through), 'folia' (leaf).
rubescens: For the reddish leaves, from Latin 'rubrus' (red), '-escens' (becoming).
sabila: For the common name for aloe in Mexico.
variegata: For the spotted leaves, from Latin 'variegatus' (variegated).
vera: The true aloe, from Latin 'vera' (in truth, real).
vulgaris: The ordinary aloe, from Latin 'vulgaris' (common).

Common names

aboes [99] [Unspecified language]
acibar [56] [Spanish] [99] [Unspecified language]
acíbar [97] [Spanish]
agave [99] [Unspecified language]
aloé [43, 104] [Portuguese] [99] [Unspecified language]
áloe [73] [Spanish]
aloë [97] [Dutch]
aloe [99] [French] [56, 97] [Spanish] [97] [Swedish] [99] [Unspecified language]
áloe de los Barbados [73] [Spanish]
aloe delle Barbados [97] [Italian]
aloe di Curacao [97] [Italian]
aloe mediterranea [97] [Italian]
aloe vera [2, 55, 109, 122] [English] [2] [French] [97] [Italian]
aloé vera [2, 97] [Portuguese]
aloé-de-Barbados [43, 104] [Portuguese]
aloé-dos-Barbados [43, 104] [Portuguese]
aloès [2, 97] [French]
aloés [2, 97] [Portuguese]
aloes [99] [Unspecified language]
aloès amer [79] [French]
aloés de Barbados [2, 97] [Portuguese]
aloès vulgaire [2, 56, 97] [French]
aloes zwyczajny [97] [Polish]
aloja [97] [Serbian]
atzavara [73] [Spanish]
azebre [99] [Unspecified language]
azebre vegetal [2, 97] [Portuguese]
babosa [43, 56, 97, 104] [Portuguese] [99] [Unspecified language]
babosa-medicinal [56] [Portuguese]
Barbados aloe [2, 32, 41, 56, 97, 99, 127] [English] [97] [Swedish]
bito-xha [32] [Zapotec]
burn plant [41] [English]
cacto-dos-aflitos [104] [Portuguese]
caraguatá [2] [Portuguese] [99] [Unspecified language]
cây aloe vera [97] [Vietnamese]
cây lô hội [97] [Vietnamese]
cây nha đam [97] [Vietnamese]
chinna kalabanda [97] [Telugu]
chirukuttali [97] [Tamil]
coastal aloe [2, 97] [English]
common aloe [99] [English]
cura-cancros [104] [Portuguese]
Curaçao aloe [2, 32, 41, 97, 99, 127] [English] [56] [Spanish]
dilang-boaia [99] [Unspecified language]
dilang-halo [99] [Unspecified language]
echte aloe [56, 97] [German]
elwa [99] [Unspecified language]

erva-azebra [43, 104] [Portuguese]
erva-babosa [2, 43, 56, 97, 104] [Portuguese] [99] [Unspecified language]
erva-que-arde [104] [Portuguese]
flor do deserto [97] [Spanish]
gavakava [97] [Shona]
ghikanvar [97] [Hindi] [99] [Unspecified language]
ghiu kumari [97] [Nepali]
ghrita kumari [97] [Bengali] [97] [Sanskrit]
ghrit-kumari [99] [Unspecified language]
guar patha [97] [Hindi]
hsiang tan [99] [Unspecified language]
humpets'k'in-ki [32] [Maya]
Indian aloe [2, 97, 99] [English]
Jaffarabad aloe [97] [English]
jaya jaya [99] [Unspecified language]
kalabanda [99] [Unspecified language]
kattalai [99] [Unspecified language]
korphad [97] [Marathi]
kumaree [99] [Unspecified language]
kumari [97] [Bengali] [97] [Malayalam] [97] [Oriya] [97] [Sanskrit]
kunvar [99] [Unspecified language]
lääkeaaloe [97] [Finnish]
lægealoe [97] [Danish]
lank'u [99] [Unspecified language]
lap'i [99] [Unspecified language]
legno aloe [97] [Italian]
linâloe [97] [Spanish]
lolisara [97] [Kannada]
loto do deserto [97] [Spanish]
lu hui [99] [Unspecified language]
maguey morado [97] [Spanish]
mazambron [79] [French]
medicinal aloe [2, 32, 41, 97] [English]
medicirula-azebre [99] [Unspecified language]
medisyneaalwyn [55, 109] [Afrikaans]
Mediterranean aloe [2, 97] [English]
no hui [99] [Unspecified language]
ödağacı [97] [Turkish]
penca sábila [97] [Spanish]
pets'in-ki [32] [Maya]
pita zabila [73] [Spanish]
pitazabila [30] [Spanish]
pitera amarelo [97] [Spanish]
planta-dos-milagres [104] [Portuguese]
planta-mistério [104] [Portuguese]
planta-que-cura [104] [Portuguese]
pohon gaharu [97] [Malay]
sábila [32, 88, 97] [Spanish] [99] [Unspecified language]
sabila [99] [Unspecified language]
sábila do penca [97] [Spanish]
sabila-pinya [99] [Unspecified language]
saqal [99] [Arabic]
sarısabır [97] [Turkish]
sarýsabýr [97] [Turkish]
sasparila [39] [Unspecified language]
sávila [32, 56, 97] [Spanish]
sawila [99] [Unspecified language]
star cactus [97] [English]
toba xa [99] [Unspecified language]
toba xha [99] [Unspecified language]
toots amarelo [97] [Spanish]
true aloe [2, 56, 97] [English]
unguentine cactus [32, 41] [English]
West Indian aloe [2, 56, 97] [English]
zabila [30, 73] [Spanish]
zábila [32, 97] [Spanish]
zábila dos toots [97] [Spanish]
zadiva [30] [Spanish]
zotollin [99] [Unspecified language]
Алое настоящее (aloe nastojaščee) [97] [Russian]
Алоэ (aloe) [97] [Russian]
Алоэ Вера (aloe vera) [97] [Russian]
تابن هولألا , ةولألا [97] [Arabic]
وبراد [97] [Persian group]
ว่านไฟไหม้ *(wan fai mai)* [97] [Thai]
ว่านหางจระเข้ *(wan hang chora khe)* [97] [Thai]
หางตะเข้ *(hang ta khe)* [97] [Thai]
アロエ *(aroe)* [97] [Japanese]
龙舌兰 [97] [Chinese]

Products

Aloe: Curaçao aloe yielding >50% water-soluble extractive[89].
Aloe Barbadensis: The residue obtained by evaporating the juice of the leaves; contains >28% hydroxyanthracene derivatives calculated as anhydrous barbaloin[16].
Aloe Vera Gel: The colourless mucilaginous gel obtained from the parenchymatous cells in the fresh leaves[128].
Barbados aloes: Concentrated and dried juice of the leaves obtained by evaporation; contains >28% hydroxyanthracene derivatives expressed as anhydrous barbaloin[18, 22, 127].

Curaçao aloes: Concentrated and dried juice of the leaves obtained by evaporation; contains >28% hydroxyanthracene derivatives expressed as anhydrous barbaloin[20, 22, 127].

Aloe verdoorniae Reynolds

Aloe verdoorniae (Photographer: SANBI, P. Joffe)

Etymology

verdoorniae: For Dr Inez C. Verdoorn (1896–1989), South African botanist and curator of the National Herbarium of the then Botanical Research Institute (now SANBI) in Pretoria, South Africa.

Common names

blou-bontaalwyn [55, 109] [Afrikaans]
blue spotted aloe [55, 109] [English]

Aloe verecunda Pole-Evans

Aloe verecunda (Photographer: J. Kirkel)

Etymology

verecunda: Because the leaves wither in winter and the plant is almost impossible to be seen, from Latin 'verecundus' (modest).

Common names

grasaalwyn [55, 109] [Afrikaans]
grass aloe [55, 109] [English]

Aloe versicolor Guillaumin var. *versicolor*

Etymology

versicolor: Perhaps for the flower colour, from Latin 'versicolor' (variously coloured).

Aloe versicolor Guillaumin var. *steffanieana* (Rauh) J.-B.Castillon & J.-P.Castillon

Synonyms

A. steffanieana Rauh

Etymology

steffanieana: For Mrs Steffanie Paulsen, German horticulturalist responsible for the Madagascar collection at the Heidelberg Botanical Garden, Germany.

versicolor: Perhaps for the flower colour, from Latin 'versicolor' (variously coloured).

Aloe veseyi Reynolds

Aloe veseyi (Photographer: SANBI, G.W. Reynolds)

Synonyms

A. enotata L.C.Leach

Etymology

enotata: For the unspotted leaves, from Latin 'enotatus' (unmarked).

veseyi: For L. Desmond E.F. Vesey-Fitzgerald (1909 or 1910–1974), British entomologist, who worked in many tropical countries, including Kenya, Tanzania and Zambia.

Aloe viguieri H.Perrier

Aloe viguieri (Photographer: G. Orlando)

Etymology

viguieri: For Prof. René Viguier (1880–1931), French botanist who collected plants in Madagascar with H. Humbert.

Aloe viridiflora Reynolds

Etymology

viridiflora: For the green flowers, from Latin 'viridis' (green), '-florus' (flowered).

Common names

groenblomaalwyn [9, 55] [Afrikaans]

Aloe vituensis Baker

Etymology

vituensis: For the erroneously presumed occurrence in the Witu region in Kenya, starting point for the expedition on which the plant was collected.

Aloe vituensis (Photographer: C.S. Björa)

Aloe vogtsii Reynolds

Aloe vogtsii (Photographer: A.W. Klopper)

Etymology

vogtsii: For Louis R. Vogts, South African administrator and cultivator of succulent plants in his garden near Pretoria, who discovered it near Louis Trichardt (Makhado) in Limpopo Province, South Africa.

Aloe volkensii Engl. subsp. *multicaulis* S.Carter & L.E.Newton

Etymology

multicaulis: For the stems clustered from the base, sometimes branched, from Latin 'multi' (many), 'caulis' (stem).

volkensii: For Prof. Georg L.A. Volkens (1855–1917), German botanist in Berlin and explorer of Mt Kilimanjaro.

Aloe volkensii Engl. subsp. *volkensii*

Aloe volkensii subsp. *volkensii* (Photographer: C.S. Björa)

Aloe vituensis

Synonyms

A. stuhlmannii Baker

Etymology

stuhlmannii: For Franz Stuhlman (1863–1927), Acting Governor of Tanganyika (Tanzania) and plant collector, who collected the type in Zanzibar.
volkensii: For Prof. Georg L.A. Volkens (1855–1917), German botanist in Berlin and explorer of Mt Kilimanjaro.

Common names

iratune [55] [Chaga]
linakha [55] [Swahili]
mradune [5] [Chaga]
os suguroi [5, 55] [Maasai]
osuguroi [55] [Maasai]

Aloe vossii Reynolds

Aloe vossii (Photographer: A.W. Klopper)

Etymology

vossii: For Harold Voss, who first collected it.

Aloe vryheidensis Groenew.

Aloe vryheidensis (Photographer: G. Nichols)

Synonyms

A. dolomitica Groenew.

Etymology

dolomitica: For the occurrence on dolomite outcrops, from Latin.
vryheidensis: For the occurrence near the town of Vryheid in KwaZulu-Natal, South Africa.

Common names

bruinaalwyn [55, 62, 95, 123] [Afrikaans]
dolomite aloe [53, 99, 124] [English]
Vryheid aloe [95] [English]
Wolkberg aloe [124] [English]
Wolkbergaalwyn [9, 55, 124] [Afrikaans]

Aloe volkensii subsp. *volkensii*

Aloe werneri J.-B.Castillon

Etymology

werneri: For Prof. Werner Rauh (1913–2000), German botanist in Heidelberg and specialist of Madagascan succulents, who discovered it in 1991 and noted its main distinguishing characters.

Aloe whitcombei Lavranos

Etymology

whitcombei: For R.P. Whitcombe of Salalah, in Oman, who first collected the plant.

Aloe wickensii Pole-Evans var. *lutea* Reynolds

Aloe wickensii var. *lutea* (Photographer: G.F. Smith)

Etymology

lutea: For the yellow flowers, from Latin 'luteus' (yellow).

wickensii: For John E. Wickens (1867–1949), English horticulturalist and plant collector in South Africa.

Aloe wickensii Pole-Evans var. *wickensii*

Aloe wickensii var. *wickensii* (Photographer: SANBI, G.W. Reynolds)

Etymology

wickensii: For John E. Wickens (1867–1949), English horticulturalist and plant collector in South Africa.

Common names

ngafane [101] [Sotho, Northern]
Wickens' aloe [55, 109] [English]
Wickens-aalwyn [109] [Afrikaans]

Aloe wildii (Reynolds) Reynolds

Synonyms

A. torrei I.Verd. & Christian var. *wildii* Reynolds

Aloe werneri

Aloe wildii (Photographer: SANBI, D.C.H. Plowes)

Etymology

torrei: For António Rocha da Torre (1904–1995), Portuguese biologist and pharmacist.
wildii: For Prof. Hiram Wild (1917–1982), British botanist and director of the National Herbarium in Harare, Zimbabwe.

Common names

Wild's small Chimanimani aloe [55, 126] [English]

Aloe wilsonii Reynolds

Etymology

wilsonii: For John G. Wilson (1927–), British agricultural officer and ecologist in the Ugandan Department of Agriculture; later lived in Kenya.

Common names

echuchukua [55, 102] [Karamojong]

Aloe woodii Lavranos & Collen.

Synonyms

A. parvicapsula Lavranos & Collen.

Etymology

parvicapsula: For the small fruits, in comparison with the allied species *Aloe woodii*, from Latin 'parvus' (small), 'capsula' (capsule).
woodii: For John R.I. Wood (1944–), British Inspector of Schools in Yemen and active amateur botanist.

Aloe wrefordii Reynolds

Etymology

wrefordii: For Herbert Wreford-Smith (1890–1962), transporter, farmer, prospector, cattle dealer and naturalist in Kenya and Uganda.

Aloe yavellana Reynolds

Etymology

yavellana: For the occurrence at Yavello, in Ethiopia.

Common names

argeesaa [40] [Oromo, West Central]
heejersaa [40] [Oromo, West Central]

Aloe yemenica J.R.I.Wood

Etymology

yemenica: For the occurrence in Yemen.

Aloe zakamisyi T.A.McCoy & Lavranos

Etymology

zakamisyi: For Mr Zakamisy, who showed the authors the way to the place where the plant was growing.

Aloe zebrina Baker

Aloe zebrina (Photographer: SANBI, P. Joffe)

Synonyms

A. bamangwatensis Schönland
A. baumii Engl. & Gilg.
A. constricta Baker
A. lugardiana Baker
A. platyphylla Baker

Etymology

bamangwatensis: For the occurrence in the country of the Bamangwatos, in Griqualand, South Africa.

baumii: For Hugo Baum (1867–1950), plant collector on the Cunene-Zambesi expedition to Angola in 1899–1900, who collected the type.

constricta: For the constricted perianth, from Latin 'constrictus' (constricted).

lugardiana: For Charlotte E. Lugard (1859–1939), painter and collector who collected the type near Botletle River in Bechuanaland (Botswana).

platyphylla: For the leaf shape, from Greek 'platys' (flat, broad), 'phyllon' (leaf).

zebrina: Referring to the spots on the leaves that often merge to form more or less regular transverse stripes or bands, as in a zebra, from Portuguese 'zebra' Latinised.

Common names

//noru [53, 55] [Naro]
//nuru [53, 55] [Naro]
/ganya [53, 55] [Unspecified language]
/gikwe [53, 55] [Naro]
||ganja [117] [Kung-Ekoka]
aloes [97] [French]
aloes pregowany [97] [Polish]
aloès tacheté [61] [French]
aloès zébré [61] [French]
aukoreb [55] [Nama]
bec de perroquet [97] [French]
bontaalwee [55, 108] [Afrikaans]
bontaalwyn [55, 59, 60, 108, 109, 122] [Afrikaans]
botaalwee [105] [Kua]
chinthembwe [82] [Nyanja] [82] [Tumbuka]
dishashanogha [55] [Mbukushu]
djimbelia [54] [Kimbundu]
edundu [53, 55] [Kwanyama]
ekundu [105] [Kua] [55] [Kwangali] [55] [Ndebele]
gilodu [55] [Nama]
gorge de perdrix [97] [French]
iandala [15] [Umbundu]
kanembe [55] [Ila] [55] [Tonga]
kanniedood aloe [97] [English]
kgopalmabalamantsi [55, 122] [Tswana]
kgope [122] [Tswana]
lichongwe [82] [Yao]
lishashankogha [55] [Diriku]
mangana [37] [Unspecified language]
mantombo [54] [Nyaneka]
n||cru [117] [Khwe] [117] [Naro]

n||uru [117] [Khwe]
njandola [15] [Unspecified language]
nl/'ho'oru [55] [Kung-Ekoka]
nllhoq'uru [55] [Nama]
nllhoq'ùrù [72] [Ju|'hoan]
ojinkalangua [54] [Kimbundu]
okandala-kasengue [15] [Unspecified language]
okandala-kazengue [54] [Umbundu]
okandole [15] [Unspecified language]
omakundu [105] [Kua] [55] [Kwangali]
otchandala-ekundu [15] [Unspecified language]
otchyandala [15] [Umbundu]
otjindombo [55] [Herero]
otyiandola [54] [Nyemba]
oviandala [15] [Umbundu]
partridge breast aloe [97] [English]
sankulu [54] [Kikongo]
spotted aloe [55, 61, 109, 122] [English]
tiger aloe [97] [English]
zebra leaf aloe [55, 56, 126] [English]
zebra-leaf aloe [56] [English]
アロエ チヨダニシキ *(aroe chiyodanishiki)* [97] [Japanese]
千代田錦 [97] [Japanese]

Aloe framesii (Photographer: A.W. Klopper)

Aloe striatula var. *caesia* (Photographer: N.R. Crouch)

PART II: NAMES FOR WHICH THE EXACT APPLICATION IS UNKNOWN

Aloe abyssinica Lam.

Synonyms

A. vulgaris Lam. var. *abyssinica* DC.

Etymology

abyssinica: For the occurrence in Abyssinia.
vulgaris: The ordinary aloe, from Latin 'vulgaris' (common).

Common names

kakamamba[5] [Zinza]

Aloe agavifolia Tod.

Etymology

agavifolia: For the leaves resembling those of an *Agave*, from Latin '-folia' (leaved).

Aloe angusta Schult. & Schult.f.

Etymology

angusta: For the narrow leaves, from Latin 'angustus' (narrow).

Aloe brownii Baker

Synonyms

A. flavescens Bouché ex A.Berger
A. nobilis Baker var. *densifolia* Baker

Etymology

brownii: For Dr Nicholas E. Brown (1849–1934), English botanist at the Royal Botanic Gardens, Kew, specialising in African succulents, especially the Mesembryanthemaceae.
densifolia: For the dense leaves, from Latin 'densus' (dense), '-folia' (leaved).
flavescens: For the yellow perianth, from Latin 'flavus' (yellow), '-escens' (becoming).
nobilis: For the size, from Latin 'nobilis' (noble).

Aloe cinnabarina Diels ex A.Berger

Etymology

cinnabarina: For the red colour of the flower, from Latin 'cinnabarinus' (cinnabar red).

Aloe consobrina Salm-Dyck

Etymology

consobrina: Probably for the relationship to other species, from Latin 'consobrinus' (cousin).

Aloe deflexidens Pillans

Etymology

deflexidens: For the deflexed teeth, from Latin 'deflexus' (deflexed), 'dens' (tooth).

Aloe dispar A.Berger

Etymology

dispar: Probably for the differences with other species of *Aloe*, from Latin 'dispar' (different, unequal).

Aloe elizae A.Berger

Etymology

elizae: For Eliza Berger, the wife of the author.

Aloe globulifera Graessn.

Etymology

globulifera: Unresolved application, from Latin 'globulus' (little ball), '-fera' (carrying).

Aloe haynaldii Tod.

Etymology

haynaldii: Unresolved application, probably for someone by the name of Haynald.

Aloe hexapetala Salm-Dyck

Synonyms

A. chloroleuca Baker
A. drepanophylla Baker
A. platylepis Baker

Etymology

chloroleuca: Probably referring to the flower, which however, is described as yellowish-white not greenish-white, from Greek 'chloros' (green), 'leucos' (white).
drepanophylla: For the leaf shape, from Greek 'drepane' (sickle), 'phyllon' (leaf).
hexapetala: For the six perianth segments, from Greek 'hexa-' (six), 'petalon' (petal).
platylepis: Application obscure, from Greek 'platys' (flat, broad), 'lepis' (scale).

Aloe leucantha A.Berger

Etymology

leucantha: For the white flowers, from Greek 'leucos' (white),'anthos' (flower).

Aloe linguiformis Medik. (nom. illegit.)

Etymology

linguiformis: For the shape of the leaves, from Latin 'lingua' (tongue), '-formis' (shaped).

Aloe longiflora Baker

Etymology

longiflora: For the overall flower size, from Latin 'longus' (long), '-florus' (flowered).

Aloe mitis A.Berger

Etymology

mitis: Application obscure, from Latin 'mitis' (mild or soft).

Aloe monteiroi Baker

Etymology

monteiroi: For Rose Monteiro (1840–1897), who collected the type specimen in Delagoa Bay (Maputo), in Mozambique.

Aloe neglecta Ten.

Etymology

neglecta: For being neglected, from Latin 'neglectus' (neglected).

Aloe obscurevirens Martinati ex Vis.

Etymology

obscurevirens: Unresolved application, from Latin 'obscurus' (indistinct, obscure), 'virens' (becoming green).

Aloe pungens A.Berger

Etymology

pungens: From Latin 'pungens' (pungent, piercing).

Aloe quinquangularis Schult.f.

Synonyms

A. pentagona Salm-Dyck

Etymology

pentagona: Unresolved application, from Greek 'penta' (five), '-gonia' (angle). It could refer to the leaves being arranged into five ranks.
quinquangularis: Unresolved application, from Latin 'quinque' (five), 'angularis' (angled).

Aloe serrulata (Aiton) Baker (possible hybrid with *A. variegata* as one parent)

Synonyms

A. pallescens Haw.
A. perfoliata L. var. *serrulata* Aiton

Etymology

pallescens: Probably for the pale flower, from Latin 'pallescens' (becoming pale).
perfoliata: For the stem passing through the leaves, i.e. the leaves are amplexicaul, from Latin 'per' (through), 'folia' (leaf).
serrulata: For the finely serrulate leaf margins, from Latin 'serrulatus' (serrulate).

Aloe sororia A.Berger

Etymology

sororia: For the relationship to other species, from Latin 'soror' (sister).

Aloe stans A.Berger

Synonyms

A. nobilis Baker (nom. illegit.)

Etymology

nobilis: Probably for the size, from Latin 'nobilis' (noble).
stans: For the erect habit, from Latin 'stans' (standing upright).

Aloe straussii A.Berger

Etymology

straussii: For Mr H. Strauss, who in 1910 sent the plant from Berlin to La Mortola, Italy, where it was described.

Aloe ×*tomlinsonii* Marloth (*A. ferox* × *A. hexapetala*)

Etymology

tomlinsonii: For Mr L.L. Tomlinson, of Delarey, Swellendam, South Africa, who discovered the plant.

Aloe vera (L.) Burm.f. var. *wratislaviensis* Kostecka-Madalska

Etymology

vera: The true aloe, from Latin 'vera' (in truth, real).
wratislaviensis: Unresolved application.

Aloe virens Haw. (possible hybrid with *A. humilis* as one parent)

Synonyms

A. virens Haw. var. *macilenta* Baker

Etymology

macilenta: For being lean, from Latin 'macilentus' (lean).
virens: For the colour of the leaves, from Latin 'virens' (becoming green).

PART III: REFERENCES

1 ABBIW, D.K. 1990. *Useful plants of Ghana: West African uses of wild and cultivated plants*. Intermediate Technology Publications, London; Royal Botanic Gardens, Kew.

2 AFOLAYAN, A.J. & ADEBOLA, P.O. 2006. *Aloe vera* (L.) Burm.f. In G.H. Schmelzer & A. Gurib-Fakim (eds), *PROTA 11(1): Medicinal plants*. PROTA (Plant Resources of Tropical Africa/Ressources végétales de l'Afrique tropicale), Wageningen, The Netherlands. Accessed online 3 February 2009: www.prota4u.org/protav8.asp?h=M26&t=Aloe_vera&p=Aloe+vera#.

3 BAKER, G.J. 1877. *Flora of Mauritius and the Seychelles—a description of the flowering plants and ferns of those islands*. Reeve, London.

4 BANDEIRA, S., BOLNICK, D. & BARBOSA, F. 2007. *Flores nativas do sul de Moçambique (Wild flowers of southern Mozambique)*. Universidade Eduardo Mondlane, Maputo.

5 BAYARD HORA, F. & GREENWAY, P.J. 1940. *Checklists of the forest trees and shrubs of the British Empire no. 5*. Imperial Forestry Institute, Oxford.

6 BEENTJE, H.J. 1994. *Kenya trees, shrubs and lianas*. National Museums of Kenya, Nairobi.

7 BIEGEL, H.M. 1979. *Rhodesian wild flowers*. Thomas Meikle Series 4. Trustees of the National Monuments and Museums of Rhodesia, Harare.

8 BIEGEL, H.M. & MAVI, S. 1972. *A Rhodesian botanical dictionary of African and English plant names*. Government Printer, Salisbury.

9 BORNMAN, H. & HARDY, D.S. 1972. *Aloes of the South African veld*. Voortrekkerpers, Johannesburg.

10 BOSCH, C.H. 2006a. *Aloe arborescens* Mill. In G.H. Schmelzer & A. Gurib-Fakim (eds), *PROTA 11(1): Medicinal plants*. PROTA (Plant Resources of Tropical Africa/Ressources Végétales de l'Afrique Tropicale), Wageningen, The Netherlands. Accessed online 3 February 2009: www.prota4u.org/protav8.asp?h=M4&t=Aloe,arborescens&p=Aloe+arborescens#.

11 BOSCH, C.H. 2006b. *Aloe ferox* Mill. In G.H. Schmelzer & A. Gurib-Fakim (eds), *PROTA 11(1): Medicinal plants*. PROTA (Plant Resources of Tropical Africa/Ressources Végétales de l'Afrique Tropicale), Wageningen, The Netherlands. Accessed online 3 February 2009: www.prota4u.org/protav8.asp?h=M26&t=Aloe_ferox&p=Aloe+ferox#.

12 BOSCH, C.H. 2006c. *Aloe lomatophylloides* Balf.f. In G.H. Schmelzer & A. Gurib-Fakim (eds), *PROTA 11(1): Medicinal plants*. PROTA (Plant Resources of Tropical Africa/Ressources Végétales de l'Afrique Tropicale), Wageningen, The Netherlands. Accessed online 3 February 2009: www.prota4u.org/protav8.asp?h=M15,M18,M34,M4,M6,M7,M9&t=Aloe,lomatophylloides&p=Aloe+lomatophylloides#.

13 BOSCH, C.H. 2006d. *Aloe nuttii* Baker. In G.H. Schmelzer & A. Gurib-Fakim (eds), *PROTA 11(1): Medicinal plants*. PROTA (Plant Resources of Tropical Africa/Ressources Végétales de l'Afrique Tropicale), Wageningen, The Netherlands. Accessed online 3 February 2009: www.prota4u.org/protav8.asp?h=M15,M18,M25,M34,M36,M4,M6,M7&t=Aloe,nuttii&p=Aloe+nuttii#.

14 BOSCH, C.H. 2006e. *Aloe wollastonii* Rendle. In G.H. Schmelzer & A. Gurib-Fakim (eds), *PROTA 11(1): Medicinal plants*. PROTA (Plant Resources of Tropical Africa/Ressources Végétales de l'Afrique Tropicale), Wageningen, The Netherlands. Accessed online 3 February 2009: www.prota4u.org/protav8.asp?h=M15,M17,M25,M34,M36,M6,M7,M9&t=Aloe,wollastonii&p=Aloe+wollastonii#.

15 BOSSARD, E. 1996. *La Medecine traditionelle au centre et a l'ouest de l'Angola*. Instituto de Investigação Científica Tropical, Lisboa.

16 BRITISH PHARMACOPOEIA. 1980a. Aloe barbadensis. *British Pharmacopoeia*. Volume I, 20–23. Her Majesty's Stationary Office, London.

17 BRITISH PHARMACOPOEIA. 1980b. Aloe capensis. *British Pharmacopoeia*. Volume I, pp. 20–26. Her Majesty's Stationary Office, London.

18 BRITISH PHARMACOPOEIA. 1980c. Barbados aloes. *British Pharmacopoeia*. Volume I, pp. 20–22. Her Majesty's Stationary Office, London.

19 BRITISH PHARMACOPOEIA. 1980d. Cape aloes. *British Pharmacopoeia*. Volume I, pp. 20–25. Her Majesty's Stationary Office, London.

20 BRITISH PHARMACOPOEIA. 1980e. Curaçao aloes. *British Pharmacopoeia*. Volume I, pp. 20–24. Her Majesty's Stationary Office, London.

21 BRITISH PHARMACOPOEIA. 1980f. Powdered aloes. *British Pharmacopoeia*. Volume I, pp. 20–21. Her Majesty's Stationary Office, London.

22 BRITISH PHARMACOPOEIA. 2009a. Barbados aloes. *British Pharmacopoeia*. Volume I. Crown Copyright, London.

23 BRITISH PHARMACOPOEIA. 2009b. Cape aloes. *British Pharmacopoeia*. Volume I, p. 3. Crown Copyright, London.

24 BRYANT, A.T. 1866. *Zulu medicine and medicine-men*. Struik, Cape Town.

25 BURKE, A. 2003. *Wild flowers of the southern Namib*. Namibia Scientific Society, Windhoek.

26 BURKE, A. 2006. *Wild flowers of the central Namib*. Namibia Scientific Society, Windhoek.

27 BURKILL, H.M. 1995. *The useful plants of West Tropical Africa*. Volume 3 families J–L. Royal Botanic Gardens, Kew.

28 CASTILLON, J.-B. 2004. Two new species and a new variety of *Aloe* (Asphodelaceae) from Madagascar. *Haseltonia* 10: 44–50.

29 CASTILLON, J.-B. & CASTILLON, J.-P. 2009. Personal communication.

30 CEBALLOS-JIMENEZ, A. 1986. *Diccionario ilustrado de los nombres vernáculos de las Plantas en Espana*. ICONA, Madrid.

31 CHRISTIE, S.J., DUTTON, R.W., HANNON, D.P., MILLER, A.G. & OAKMAN, N.A. 2005. *Aloe jawiyon*, a new species from Soqotra (Yemen). *Bradleya* 23: 23–30.

32 CONABIO. 2009. *Comisión Nacional para el Conocomiento y Uso de la Biodiverisdad*. Accessed online: www.conabio.gob.mx/malezasdemexico/asphodelaceae/aloe-vera/fichas/ficha.htm.

33 CRAVEN, P. & MARAIS, C. 1986. *Damaraland Flora. Spitzkoppe, Brandberg, Twyfelfontein*. Gamsberg Macmillan, Windhoek.

34 CRAVEN, P. & MARAIS, C. (1986). *Namib Flora. Swakopmund to the giant Welwitschias via Coanikontes*. Gamsberg Macmillan, Windhoek.

35 CUNNINGHAM, A.B. 1988. *An investigation of the herbal medicine trade in Natal/KwaZulu*. Investigational report no. 29. Institute of Natural Resources, Pietermaritzburg.

36 CURTIS, B. & MANNHEIMER, C. 2005. *Tree atlas of Namibia*. National Botanical Research Institute, Windhoek.

37 DA SILVA, M.C., IZIDINE, S. & AMUDE, A.B. 2004. *A preliminary checklist of the vascular plants of Mozambique—Catálogo provisório das plantas superiores de Moçambique*. *Southern African Botanical Diversity Network Report No*. 30. SABONET, Pretoria.

38 DALZIEL, J.M. 1937. *The useful plants of West Tropical Africa*. Crown Agents for the Colonies, London.

39 DEIGHTON, F.C. 1957. *Vernacular botanical vocabulary for Sierra Leone*. Crown Agents for Oversea Governments and Administrations, London.

40 EDWARDS, S., DEMISSEW, S. & HEDBERG, I. 1996. *Flora of Ethiopia and Eritrea*. Volume 6. The National Herbarium, Addis Ababa University, Addis Ababa; Department of Systematic Botany, Uppsala University, Uppsala.

41 EFLORAS. 2009. *eFloras.* Accessed online: www.efloras.org/florataxon.aspx?flora_id=1&taxon_id=200027555.

42 EGGLI, U. & NEWTON, L.E. 2004. *Etymological dictionary of succulent plant names*. Springer-Verlag, Berlin.

43 FERNANDES, F.M. & CARVALHO, L.M. 2003. *Portugal botânico de A a Z—Plantas Portuguesas e exóticas*. Lidel-Edições Técnicas, Lousã.

44 FOSBERG, F.R. & RENVOIZE, S.A. 1980. *The Flora of Aldabra and neighbouring islands*. Kew Bulletin additional series VII. Royal Botanic Gardens, Kew.

45 GELFAND, M., MAVI, S., DRUMMOND, R.B. & NDEMERA, B. 1985. *The traditional medical practitioner in Zimbabwe: his principles of practice and pharmacopoeia*. Mambo Press, Gweru.

46 GERSTNER, J. 1938a. A preliminary check list of Zulu names of plants. *Bantu Studies* 12: 215–236.

47 GERSTNER, J. 1938b. A preliminary check list of Zulu names of plants with short notes (continued). *Bantu Studies* 12: 321–342.

48 GERSTNER, J. 1939a. A preliminary check list of Zulu names of plants with short notes (continued). *Bantu Studies* 13: 49–64.

49 GERSTNER, J. 1939b. A preliminary check list of Zulu names of plants with short notes (continued). *Bantu Studies* 13: 131–149.

50 GERSTNER, J. 1939c. A preliminary check list of Zulu names of plants with short notes (continued). *Bantu Studies* 13: 307–326.

51 GERSTNER, J. 1941a. A preliminary check list of Zulu names of plants with short notes (continued). *Bantu Studies* 15: 277–301.

52 GERSTNER, J. 1941b. A preliminary check list of Zulu names of plants with short notes (continued). *Bantu Studies* 15: 369–383.

53 GLEN, H.F. & HARDY, D.S. 2000. Aloaceae (First part): *Aloe*. In G. Germishuizen (ed.), *Flora of southern Africa*. Volume 5 part 1 fascicle 1. National Botanical Institute, Pretoria.

54 GOSSWEILER, J. 1953. *Nomes indígenas de plantas de Angola. Agronomia Angolana* 7. Imprensa Nacional, Luanda.

55 GRACE, O.M. 2009. Database of ethnobotanical uses of the genus *Aloe*. O.M. Grace, London.

56 GRIN. 2009. *Germplasm resources information network—GRIN.* National Germplasm Resources Laboratory, Beltsville, SA. Accessed online 5 February 2009: www.ars-grin.gov/cgi-bin/npgs/html/taxon.pl?2510.

57 HAHN, N. 1994. *Tree list of the Soutpansberg*. Fantique, Hatfield.

58 JACOT GUILLARMOD, A. 1971. *Flora of Lesotho.* J. Cramer, Lehle.

59 JANKOWITZ, W.A. 1975. *Aloes of South West Africa*. Division of Nature Conservation and Tourism, Windhoek.

60 JANKOWITZ, W.A. 2009. Personal communication.

61 JANSEN, P.C.M. 2006. *Aloe zebrina* Baker. In P.C.M. Jansen & D. Cardon (eds), *PROTA 3: Dyes and tannins*. PROTA (Plant Resources of Tropical Africa/Ressources Végétales de l'Afrique Tropicale), Wageningen, The Netherlands. Accessed online 3 February 2009: www.prota4u.org/protav8.asp?h=M11,M12,M15,M17,M18,M25,M26,M34,M4,M5,M6,M7,M9&t=Aloe,zebrina,aloe&p=Aloe+zebrina#.

62 JEPPE, B. 1969. *South African Aloes*. Purnell, Cape Town.

63 JORDAAN, P.G. 2009. Data from herbarium specimen labels, Stellenbosch University Herbarium.

64 KEMP, E.S. 1983. *Trees of Swaziland*. Occasional paper no. 3. Swaziland National Trust Commission, Lobamba.

65 KILLICK, D. 1990. *A field guide to the Flora of the Natal Drakensberg*. Jonathan Ball & Ad. Donker, Johannesburg.

66 KLOSS, J. 1982. *A human interest story of health and restoration to be found in herb, root, and bark*. Back to Eden Books, Loma Linda, California.

67 LAMARCK, J.B.A.P.M. DE. 1783. *Encyclopedie Methodique, Botanique. Volume* I (1), 1–748. Pancouke, Paris.

68 LANE, S.S. 2004. *A field guide to the aloes of Malawi*. Umdaus, Hatfield.

69 LANE, S.S. 2009. Personal communication.

70 LANE, S.S., SMITH, G.F. & STEYN, E.M.A. 2003. Validation and amendment of the name *Aloe canii*. *Aloe* 40: 55.

71 LE ROUX, A. 2005. *Namakwaland. South African Wild Flower Guide* 1. Botanical Society of South Africa, Cape Town.

72 LEFFERS, A. 2003. *Gemsbok bean and Kalahari truffle*. Gamsberg MacMillan Publishers, Windhoek.

73 LÓPEZ-GONZÁLEZ, G. 2007. *Guía de los árboles y arbustos de la peninsula Ibérica y Islas Baleares*. Ediciones Mundi-Prensa, Madrid.

74 MABOGO, D.E.N. 1990. *The ethnobotany of the Vhavenda*. University of Pretoria, Pretoria.

75 MANNHEIMER, C., MAGGS-KÖLLING, G., KOLBERG, H. & RÜGHEIMER, S. 2003. *Wildflowers of the southern Namib*. National Botanical Research Institute, Windhoek.

76 MANNING, J. 2001. *Eastern Cape. South African Wild Flower Guide* 11. Botanical Society of South Africa, Cape Town.

77 MANNING, J. & GOLDBLATT, P. 1996. *West Coast. South African Wild Flower Guide* 7. Botanical Society of South Africa, Cape Town.

78 MANNING, J. & GOLDBLATT, P. 1997. *Nieuwoudtville, Bokkeveld Plateau and Hantam. South African Wild Flower Guide* 9. Botanical Society of South Africa, Cape Town.

79 MARAIS, W. 1978. *Aloe* L. In J. Bosser, T. Cadet, H.R. Julien & W. Marais (eds), *Flore des Mascareignes, La Réunion, Maurice, Rodrigues.* Volume 183. Liliacées, pp. 9–10. Sugar Industry Research Institute, Mauritius.

80 MARAIS, W. 1978. *Lomatophyllum* Willd. In J. Bosser, T. Cadet, H.R. Julien & W. Marais (eds), *Flore des Mascareignes, La Réunion, Maurice, Rodrigues.* Volume 183. Liliacées, pp. 10–14. Sugar Industry Research Institute, Mauritius.

81 MORIARTY, A. 1982. *Outeniqua, Tsitsikamma and Eastern Little Karoo. South African Wild Flower Guide* 2. Botanical Society of South Africa, Cape Town.

82 MSEKANDIANA, G. 2009. Personal communication.

83 MUSTART, P., COWLING, R. & ALBERTYN, J. 1997. *Southern Overberg. South African Wild Flower Guide* 8. Botanical Society of South Africa, Cape Town.

84 NEWINGER, H.D. 1996. *African ethnobotany: poisons and drugs, chemistry, pharmacology, toxicology*. Chapman & Hall, London.

85 NEWTON, L.E. 2006. *Aloe lateritia* Engl. In G.H. Schmelzer & A. Gurib-Fakim (eds), *PROTA 11(1): Medicinal plants*. PROTA (Plant Resources of Tropical Africa/Ressources Végétales de l'Afrique Tropicale), Wageningen, The Netherlands. Accessed online 03 February 2009: www.prota4u.org/protav8.asp?h=M12,M15,M16,M18,M20,M21,M23,M24,M25,M26,M27,M34,M36,M4,M6,M7,M8,M9&t=Aloe,lateritia&p=Aloe+lateritia#.

86 ONDERSTALL, J. 1984. *Transvaal Lowveld and Escarpment. South African Wild Flower Guide* 4. Botanical Society of South Africa, Cape Town.

87 O'NEIL, M.J. 2009. Aloe monograph 0000305 2564. In M.J. O'Neil (ed.), *The Merck Index: an encyclopedia of chemicals, drugs, and biologicals*. 14th edition. Merck, New Jersey.

88 PARDO, O. 2002. *Etnobotánica de algunas cactáceas y suculentas del Perú* 5. *Chloris Chilensis*. Accessed online: www.chlorischile.cl.

89 PARFITT, K. 1999. Aloes. In K. Parfitt (ed.), *Martindale: the complete drug reference*. 32nd edition. Pharmaceutical Press, Taunton.

90 PERRIER DE LA BATHIE, H. 1926. Les *Lomatophyllum* et les *Aloe* de Madagascar. *Mémoires de la Société Linnéenne de Normandie* 1: 1–59.

91 PERRIER DE LA BATHIE, H. 1938. Liliacées: Aloineae: *Aloe*. *Flore de Madagascar* 40: 77–112. Imprimerie Officielle, Tananarive, Madagascar.

92 PHARMACEUTICAL CODEX. 1979. *The Pharmaceutical Codex incorporating the British Pharmaceutical Codex*. Pharmaceutical Press, London.

93 PIENAAR, A. 2008. *Kruidjie roer my nie: die antieke helingskuns van die Karoo-veld*. Umuzi, Cape Town.

94 POOLEY, E. 1993. *The complete field guide to trees of Natal Zululand & Transkei*. Natal Flora Publications Trust, Durban.

95 POOLEY, E. 1998. *A field guide to wildflowers of KwaZulu-Natal and the eastern regions*. Natal Flora Publications Trust, Durban.

96 POOLEY, E. 2003. *Mountain flowers. A field guide to the flora of the Drakensberg and Lesotho*. Natal Flora Publications Trust, Durban.

97 PORCHER, M.H. 1995. *Multilingual multiscript plant name database*. University of Melbourne, Melbourne, Australia. Accessed online: www.plantnames.unimelb.edu.au/Sorting/Aloe.html.

98 POWRIE, L. 2004. Common names of Karoo plants. *Strelitzia* 16. South African National Biodiversity Institute, Pretoria.

99 QUATTROCCHI, U. 2000. *CRC world dictionary of plant names: common names, scientific names, eponyms, synonyms, and etymology*. CRC Press, Boca Raton.

100 RAKOTOARISOA, S.E. 2009. Personal communication.

101 REYNOLDS, G.W. 1950. *The aloes of South Africa*. Aloes of South Africa Book Fund, Johannesburg.

102 REYNOLDS, G.W. 1966. *The aloes of Tropical Africa and Madagascar*. Aloes Book Fund, Mbabane.

103 REYNOLDS, P. & CRAWFORD COUSINS, G. 1991. *Lwaano lwanyika (Tonga book of the Earth)*. Colleen Crawford Cousins, Harare.

104 ROCHA, F. 1996. *Nomes vulgares de plantas existentes em Portugal*. Direcção Geral de Protecção das Culturas Geral de Protecção das Culturas, Lisboa.

105 RODIN, J. 1985. *The ethnobotany of the Kwanyama Ovambos*. Missouri Botanic Garden, St. Louis.

106 SHEARING, D. 1994. *Karoo. South African Wild Flower Guide* 6. Botanical Society of South Africa, Cape Town.

107 SMITH, A. 1888. *A contribution to the South African materia medica, chiefly from plants in use among the natives*. 2nd edition. Lovedale.

108 SMITH, C.A. 1966. *Common names of South African plants. Botanical Survey Memoir no.* 35. Department of Agricultural and Technical Services, Pretoria.

109 SMITH, G.F. 2003. *First field guide to aloes of Southern Africa*. Struik, Cape Town.

110 SMITH, G.F. 2005. *The fascinating world of the grass aloes of South Africa.* In C. Craib (ed), *Grass aloes of the South African veld*, pp. viii–ix. Umdaus Press, Hatfield.

111 SMITH, G.F. & CROUCH, N.R. 2006a. *Asphodelaceae*: *Aloe vanrooyenii*: a distinctive new maculate *Aloe* from KwaZulu-Natal, South Africa. *Bothalia* 36: 73–75.

112 SMITH, G.F. & CROUCH, N.R. 2006b. *Asphodelaceae*. Corrections to the eponymy and geographical distribution of *Aloe vanrooyenii*. *Bothalia* 36: 174.

113 SMITH, G.F., CROUCH, N.R. & CONDY, G. 1999. *Aloe pruinosa*. *Flowering Plants of Africa* 56: t. 2141.

114 SMITH, G.F., STEYN, E.M.A. & CROUCH, N.R. 2005. *Aloe affinis*. *Aloaceae*. *Curtis's Botanical Magazine* 22: 95–99.

115 SMITH, G.F. & VAN WYK, A.E. 1989. Biographical notes on James Bowie and the discovery of *Aloe bowiea* Schult. & J.H. Schult. (Alooideae: Asphodelaceae). *Taxon* 38: 557–568.

116 SMITH, G.F., VAN WYK, B-E. & GLEN, H.F. 1994. *Aloe barberae* to replace *A. bainesii*. *Bothalia* 24: 34–35.

117 STORY, R. 1958. *Some plants used by the Bushman in obtaining food and water*. *Botanical Survey Memoir* no. 30. Department of Agricultural and Technical Services, Pretoria.

118 TREDGOLD, M.H. 1990. *Food plants of Zimbabwe with old and new ways of preparation*. Mambo Press, Harare.

119 TYISO, S. & BHAT, R.B. 1998. Medicinal plants used for child welfare in the Transkei region of the Eastern Cape (South Africa). *Angewandte Botanik* 72: 92–98.

120 VAN ROOYEN, G. & STEYN, H. 1999. *Cederberg, Clanwilliam & Biedouw Valley. South African Wild Flower Guide* 10. Botanical Society of South Africa, Cape Town.

121 VAN WYK, A.E. & MALAN, S. 1997. *Field guide to the wild flowers of the Highveld*. Struik, Cape Town.

122 VAN WYK, B-E. & GERICKE, N. 2003. *People's plants*. Briza, Pretoria.

123 VAN WYK, B-E. & SMITH, G.F. 1996. *Guide to the aloes of South Africa*. Briza, Pretoria.

124 VON BREITENBACH, F. 1990. *National list of indigenous trees.* Dendrological Foundation, South Africa.

125 WELLS, M.J., BASINHAS, A.A., JOFFE, H., ENGELBRECHT, V.M., HARDING, G. & STIRTON, C.H. 1986. *A catalogue of problem plants in southern Africa incorporating the national weed list of southern Africa*. *Memoirs of the Botanical Survey of South Africa* 53. Botanical Research Institute, Pretoria.

126 WEST, O. 1974. *A field guide to the aloes of Rhodesia*. Longman Rhodesia, Salisbury.

127 WHO. 1999a. *Aloe. WHO Monographs on Selected Medicinal Plants*. Volume I: 33–42. World Health Organization, Geneva.

128 WHO. 1999b. *Aloe vera* gel. *WHO Monographs on Selected Medicinal Plants*. Volume I: 43–49. World Health Organization, Geneva.

129 WILD, H., BIEGEL, H.M. & MAVI, S. 1975. *A Rhodesian botanical dictionary of African and English plant names*. Government Printer, Salisbury.

130 WOOD, J.M. 1915. *List of trees, shrubs, and a selection of herbaceous plants grown in the Durban Municipal Botanic Gardens, with a few remarks on each*. Bennett & Davis, Durban.

PART IV: LISTS

Synonyms

Synonym	Accepted name
Aloe abyssinica A.Berger (nom. illegit.)	*Aloe elegans* Tod.
Aloe abyssinica Lam. var. *peacockii* Baker.	*Aloe elegans* Tod.
Aloe abyssinica Lam. var. *percrassa* (Tod.) Baker (nom. illegit.)	*Aloe percrassa* Tod.
Aloe abyssinica Salm-Dyck (nom. illegit.)	*Aloe camperi* Schweinf.
Aloe acuminata Haw.	*Aloe humilis* (L.) Mill.
Aloe acuminata Haw. var. *major* Salm-Dyck.	*Aloe humilis* (L.) Mill.
Aloe acutissima H.Perrier var. *itampoloana* J.-B.Castillon	*Aloe acutissima* H.Perrier subsp. *itampolensis* Rebmann
Aloe aethiopica (Schweinf.) A.Berger	*Aloe elegans* Tod.
Aloe affinis Pole-Evans (nom. illegit.)	*Aloe parvibracteata* Schönland
Aloe africana Mill. var. *angustior* Haw.	*Aloe africana* Mill.
Aloe africana Mill. var. *latifolia* Haw.	*Aloe africana* Mill.
Aloe agrophila Reynolds	*Aloe ecklonis* Salm-Dyck
Aloe albispina Haw.	*Aloe perfoliata* L.
Aloe albocincta Haw.	*Aloe striata* Haw.
Aloe albopicta A.Berger	*Aloe camperi* Schweinf.
Aloe altimatsiatrae J.-B.Castillon	*Aloe fievetii* Reynolds var. *altimatsiatrae* (J.-B.Castillon) J.-B.Castillon
Aloe amaniensis A.Berger	*Aloe lateritia* Engl. var. *lateritia*
Aloe amoena Pillans	*Aloe framesii* L.Bolus
Aloe andohahelensis J.-B.Castillon	*Aloe teissieri* Lavranos
Aloe angiensis De Wild.	*Aloe macrocarpa* Tod. subsp. *wollastonii* (Rendle) Wabuyele
Aloe angiensis De Wild. var. *kitaliensis* Reynolds	*Aloe macrocarpa* Tod. subsp. *wollastonii* (Rendle) Wabuyele
Aloe angustifolia Groenew. (nom. illegit.).	*Aloe vandermerwei* Reynolds
Aloe angustifolia Haw.	*Aloe africana* Mill.
Aloe ankaranensis Rauh & Mangelsdorff	*Aloe prostrata* (H.Perrier) L.E.Newton & G.D.Rowley
Aloe ×*antoninii* A.Berger (nom. superfl.).	*Aloe* ×*hanburyi* A.Borzí
Aloe arabica Lam..	*Aloe pendens* Forssk.
Aloe arabica Salm-Dyck (nom. illegit.).	*Aloe microstigma* Salm-Dyck
Aloe arborea Medik.	*Aloe arborescens* Mill. subsp. *arborescens*

Aloe arborescens Mill. var. *frutescens* (Salm-Dyck) Link . . . . . *Aloe arborescens* Mill. subsp. *arborescens*

Aloe arborescens Mill. var. *milleri* A.Berger . . . . *Aloe arborescens* Mill. subsp. *arborescens*

Aloe arborescens Mill. var. *natalensis* (J.M.Wood & M.S.Evans) A.Berger. . . . *Aloe arborescens* Mill. subsp. *arborescens*

Aloe arborescens Mill. var. *pachythyrsa* A.Berger . . . . *Aloe arborescens* Mill. subsp. *arborescens*

Aloe arborescens Mill. var. *ucriae* (A.Terracc.) A.Berger. . . . . *Aloe* ×*ucriae* A.Terrac.

Aloe arborescens Mill. var. *viridifolia* A.Berger . . . . *Aloe arborescens* Mill. subsp. *arborescens*

Aloe aristata Haw. var. *leiophylla* Baker. . . . . *Aloe aristata* Haw.

Aloe aristata Haw. var. *parvifolia* Baker. . . . . *Aloe aristata* Haw.

Aloe atherstonei Baker . . . . *Aloe pluridens* Haw.

Aloe audhalica Lavranos & D.S.Hardy . . . . *Aloe vacillans* Forssk.

Aloe aurantiaca Baker . . . . *Aloe striatula* Haw. var. *striatula*

Aloe ausana Dinter . . . . *Aloe variegata* L.

Aloe bainesii Dyer. . . . . *Aloe barberae* Dyer

Aloe bainesii Dyer var. *barberae* (Dyer) Baker . . . . *Aloe barberae* Dyer

Aloe bakeri Hook.f. ex Baker (nom. illegit.) . . . . *Aloe percrassa* Tod.

Aloe bamangwatensis Schönland . . . . *Aloe zebrina* Baker

Aloe barbadensis Mill.. . . . . *Aloe vera* (L.) Burm.f.

Aloe barbadensis Mill. var. *chinensis* Haw. . . . . *Aloe vera* (L.) Burm.f.

Aloe barteri Baker, p.p. . . . . *Aloe buettneri* A.Berger

Aloe barteri Baker, p.p. (leaf only) . . . . *Aloe schweinfurthii* Baker

Aloe barteri Baker var. *lutea* A.Chev.. . . . . *Aloe schweinfurthii* Baker

Aloe baumii Engl. & Gilg.. . . . . *Aloe zebrina* Baker

Aloe beniensis De Wild.. . . . . *Aloe dawei* A.Berger

Aloe bequaertii De Wild. . . . . . *Aloe macrocarpa* Tod. subsp. *wollastonii* (Rendle) Wabuyele

Aloe berhana Reynolds . . . . *Aloe debrana* Christian

Aloe boastii Letty . . . . *Aloe chortolirioides* A.Berger var. *chortolirioides*

Aloe boehmii Engl. . . . . *Aloe lateritia* Engl. var. *lateritia*

Aloe bolusii Baker. . . . . *Aloe africana* Mill.

Aloe boranensis Cufod. . . . . *Aloe otallensis* Baker

Aloe borziana A.Terracc. . . . . *Aloe macrocarpa* Tod. subsp. *macrocarpa*

Aloe bourea Schult. & Schult.f.. . . . . *Aloe bowiea* Schult. & Schult.f.

Aloe boylei Baker subsp. *major* Hilliard & B.L.Burtt . . . . *Aloe boylei* Baker

Aloe brevifolia (Aiton) Haw. (nom. illegit.) . . . . *Aloe distans* Haw.

Aloe brevifolia Mill. var. *postgenita* (Schult. & Schult.f.) Baker. . . . . *Aloe brevifolia* Mill. var. *brevifolia*

Aloe brevifolia Mill. var. *serra* (DC.) A.Berger . . . .*Aloe brevifolia* Mill. var. *depressa* (Haw.) Baker
Aloe brevioribus Mill. . . . . *Aloe brevifolia* Mill. var. *brevifolia*
Aloe brunneo-punctata Engl. & Gilg . . . . *Aloe nuttii* Baker
Aloe brunnthaleri A.Berger ex Cammerl. . . . . *Aloe microstigma* Salm-Dyck

Aloe ×*caesia* Salm-Dyck . . . . *Aloe* ×*principis* (Haw.) Stearn
Aloe campylosiphon A.Berger. . . . . *Aloe lateritia* Engl. var. *lateritia*
Aloe candelabrum Engl. & Drude (nom. illegit.) . . . . *Aloe thraskii* Baker
Aloe capitata Baker var. *cipolinicola* H.Perrier . . . .*Aloe cipolinicola* (H.Perrier) J.-B.Castillon & J.-P.Castillon
Aloe capitata Baker var. *gneissicola* H.Perrier. . . . . *Aloe gneissicola* (H.Perrier) J.-B.Castillon & J.-P.Castillon
Aloe capitata Baker var. *trachyticola* H.Perrier . . . . . . *Aloe trachyticola* (H.Perrier) Reynolds var. *trachyticola*
Aloe caricina A.Berger. . . . . *Aloe myriacantha* (Haw.) Schult. & Schult.f.
Aloe carowii Reynolds . . . . *Aloe sladeniana* Pole-Evans
Aloe cascadensis Kuntze . . . . *Aloe striatula* Haw. var. *striatula*
Aloe cephalophora Lavranos & Collen.. . . . . *Aloe sheilae* Lavranos
Aloe cernua Tod.. . . . . *Aloe capitata* Baker var. *capitata*
Aloe chabaudii Schönland var. *verekeri* Christian. . . .*Aloe chabaudii* Schönland var. *chabaudii*
Aloe chimanimaniensis Christian . . . . *Aloe swynnertonii* Rendle
Aloe chinensis Steud. ex Baker. . . . . *Aloe vera* (L.) Burm.f.
Aloe chloroleuca Baker . . . .*Aloe hexapetala* Salm-Dyck (exact application unknown)
Aloe chortolirioides A.Berger var. *boastii* (Letty) Reynolds . . . .*Aloe chortolirioides* A.Berger var. *chortolirioides*
Aloe ciliaris Haw. var. *flanaganii* Schönland. . . . .*Aloe ciliaris* Haw. var. *ciliaris*
Aloe commelinii Willd. . . . . *Aloe perfoliata* L.
Aloe commutata Tod. . . . . *Aloe maculata* All.
Aloe commutata Tod. var. *bicolor* (Baker) A.Berger. . . . . *Aloe maculata* All.
Aloe comosibracteata Reynolds . . . . *Aloe davyana* Schönland
Aloe compacta Reynolds . . . . *Aloe macrosiphon* Baker
Aloe compressa H.Perrier var. *rugo-squamosa* H.Perrier . . . *Aloe rugo-squamosa* (H.Perrier) J.-B.Castillon & J.-P.Castillon
Aloe concinna Baker (nom. illegit.) . . . .*Aloe squarrosa* Baker ex Balf.f.
Aloe congolensis De Wild. & T.Durand. . . . .*Aloe buettneri* A.Berger
Aloe constricta Baker . . . . *Aloe zebrina* Baker
Aloe contigua (H.Perrier) Reynolds . . . . *Aloe imalotensis* Reynolds var. *imalotensis*
Aloe corbisieri De Wild. . . . . *Aloe nuttii* Baker
Aloe cremersii Lavranos . . . .*Aloe ibitiensis* H.Perrier
Aloe cyrillei J.-B.Castillon . . . .*Aloe ibitiensis* H.Perrier

Aloe davyana Schönland var. *subolifera* Groenew. . . . . *Aloe davyana* Schönland

Aloe decora Schönland . . . . *Aloe claviflora* Burch.

Aloe decurvidens Groenew. . . . . *Aloe parvibracteata* Schönland

Aloe defalcata Chiov. . . . . *Aloe microdonta* Chiov.

Aloe deltoideodonta Baker subsp. *esomonyensis* Rebmann . . . . . *Aloe deltoideodonta* Baker var. *ruffingiana* (Rauh & Petignat) J.-B.Castillon & J.-P.Castillon

Aloe deltoideodonta Baker var. *contigua* H.Perrier . . . . *Aloe imalotensis* Reynolds var. *imalotensis*

Aloe dependens Steud. . . . . *Aloe pendens* Forssk.

Aloe depressa Haw. . . . . *Aloe brevifolia* Mill. var. *depressa* (Haw.) Baker

Aloe dhalensis Lavranos . . . . *Aloe vacillans* Forssk.

Aloe dichotoma Masson subsp. *pillansii* (L.Guthrie) Zonn.. . . . . *Aloe pillansii* L.Guthrie

Aloe dichotoma Masson subsp. *ramosissima* (Pillans) Zonn. . . . . . *Aloe ramosissima* Pillans

Aloe dichotoma Masson var. *montana* (Schinz) A.Berger . . . . *Aloe dichotoma* Masson

Aloe dichotoma Masson var. *ramosissima* (Pillans) Glen & D.S.Hardy. . . . *Aloe ramosissima* Pillans

Aloe disticha L. var. *plicatilis* L. . . . . *Aloe plicatilis* (L.) Mill.

Aloe disticha Mill. (nom. illegit.). . . . . *Aloe maculata* All.

Aloe doei Lavranos . . . . *Aloe splendens* Lavranos

Aloe doei Lavranos var. *lavranosii* Marn.-Lap.. . . . . *Aloe splendens* Lavranos

Aloe dolomitica Groenew. . . . . *Aloe vryheidensis* Groenew.

Aloe dorsalis Haw. . . . . *Aloe lineata* (Aiton) Haw. var. *lineata*

Aloe drepanophylla Baker . . . . . . *Aloe hexapetala* Salm-Dyck (exact application unknown)

Aloe dumetorum B.Mathew & Brandham . . . . *Aloe ellenbeckii* A.Berger

Aloe echinata Willd. . . . . *Aloe humilis* (L.) Mill.

Aloe echinata Willd. var. *minor* Salm-Dyck . . . . *Aloe humilis* (L.) Mill.

Aloe edulis A.Chev. ex Hutch. & Dalziel. . . . . *Aloe macrocarpa* Tod. subsp. *macrocarpa*

Aloe ellenbergeri Guillaumin . . . . *Aloe aristata* Haw.

Aloe elongata Murray . . . . *Aloe vera* (L.) Burm.f.

Aloe engleri A.Berger . . . . *Aloe secundiflora* Engl. var. *secundiflora*

Aloe enotata L.C.Leach . . . . *Aloe veseyi* Reynolds

Aloe eru A.Berger . . . . *Aloe camperi* Schweinf.

Aloe eru A.Berger var. *cornuta* A.Berger. . . . . *Aloe camperi* Schweinf.

Aloe eru A.Berger var. *hookeri* A.Berger. . . . . *Aloe adigratana* Reynolds

Aloe estevei Rebmann. . . . . *Aloe fievetii* Reynolds var. *altimatsiatrae* (J.-B.Castillon) J.-B.Castillon

Aloe eylesii Christian. . . . . *Aloe rhodesiana* Rendle

Aloe ferox Mill. var. *erythrocarpa* A.Berger. . . . . *Aloe ferox* Mill.
Aloe ferox Mill. var. *galpinii* (Baker) Reynolds . . . . *Aloe ferox* Mill.
Aloe ferox Mill. var. *hanburyi* Baker . . . . *Aloe ferox* Mill.
Aloe ferox Mill. var. *incurva* Baker . . . . *Aloe ferox* Mill.
Aloe ferox Mill. var. *subferox* (Spreng.) Baker. . . . . *Aloe ferox* Mill.
Aloe flabelliformis Salisb. . . . . *Aloe plicatilis* (L.) Mill.
Aloe flava Pers. . . . . *Aloe vera* (L.) Burm.f.
Aloe flavescens Bouché ex A.Berger . . . . . . *Aloe brownii* Baker (exact application unknown)
Aloe flavispina Haw. . . . . *Aloe perfoliata* L.
Aloe floramaculata Christian . . . . *Aloe secundiflora* Engl. var. *secundiflora*
Aloe fontainei Rebmann . . . . *Aloe mandotoensis* J.-B.Castillon
Aloe frutescens Salm-Dyck . . . . *Aloe arborescens* Mill. subsp. *arborescens*
Aloe fruticosa Lam.. . . . . *Aloe arborescens* Mill. subsp. *arborescens*
Aloe fulgens Tod. . . . . *Aloe arborescens* Mill. subsp. *arborescens*

Aloe galpinii Baker . . . . *Aloe ferox* Mill.
Aloe gasterioides Baker . . . . *Aloe maculata* All.
Aloe gillilandii Reynolds . . . . *Aloe sabaea* Schweinf.
Aloe glauca Mill. var. *elatior* Salm-Dyck . . . . *Aloe glauca* Mill.
Aloe glauca Mill. var. *humilior* Salm-Dyck . . . . *Aloe glauca* Mill.
Aloe glauca Mill. var. *major* Haw.. . . . . *Aloe glauca* Mill.
Aloe glauca Mill. var. *minor* Haw.. . . . . *Aloe glauca* Mill.
Aloe glauca Mill. var. *muricata* (Schult.) Baker. . . . . *Aloe glauca* Mill.
Aloe glauca Mill. var. *spinosior* Haw.. . . . . *Aloe glauca* Mill.
Aloe gloveri Reynolds & P.R.O.Bally. . . . . *Aloe hildebrandtii* Baker
Aloe gracilis Baker (nom. illegit.) . . . . *Aloe commixta* A.Berger
Aloe gracilis Haw. var. *decumbens* Reynolds . . . . . *Aloe decumbens* (Reynolds) van Jaarsv.
Aloe grahamii Schönland . . . . *Aloe maculata* All.
Aloe graminicola Reynolds. . . . . . . *Aloe lateritia* Engl. var. *graminicola* (Reynolds) S.Carter
Aloe graminifolia A.Berger . . . . *Aloe myriacantha* (Haw.) Schult. & Schult.f.
Aloe grandidentata Tod. (nom. illegit.). . . . . *Aloe maculata* All.
Aloe greatheadii Schönland var. *davyana* (Schönland) Glen & D.S.Hardy. . . . . *Aloe davyana* Schönland
Aloe greenwayi Reynolds . . . . *Aloe leptosiphon* A.Berger

Aloe hanburiana Naudin . . . . *Aloe striata* Haw.
Aloe harmsii A.Berger . . . . *Aloe dorotheae* A.Berger
Aloe hereroensis Engl. var. *orpeniae* (Schönland) A.Berger . . . . . *Aloe hereroensis* Engl. var. *hereroensis*

Aloe hijazensis Lavranos & Collen. . . . . *Aloe castellorum* J.R.I.Wood
Aloe horrida Haw. . . . .*Aloe ferox* Mill.
Aloe howmanii Reynolds . . . . . .*Aloe hazeliana* Reynolds var. *howmanii* (Reynolds) S.Carter
Aloe humilis (L.) Mill. var. *acuminata* (Haw.) Baker . . . . *Aloe humilis* (L.) Mill.
Aloe humilis (L.) Mill. var. *candollei* Baker. . . . . *Aloe humilis* (L.) Mill.
Aloe humilis (L.) Mill. var. *echinata* (Willd.) Baker subvar. *minor* Salm-Dyck. . . .*Aloe humilis* (L.) Mill.
Aloe humilis (L.) Mill. var. *echinata* (Willd.) Baker . . . . *Aloe humilis* (L.) Mill.
Aloe humilis (L.) Mill. var. *incurva* Haw. subvar. *minor* (Salm-Dyck) A.Berger. . .*Aloe humilis* (L.) Mill.
Aloe humilis (L.) Mill. var. *incurva* Haw. . . . . *Aloe humilis* (L.) Mill.
Aloe humilis (L.) Mill. var. *macilenta* Baker. *Aloe humilis* (L.) Mill.
Aloe humilis (L.) Mill. var. *suberecta* (Aiton) Baker subvar. *semiguttata* Haw. . . .*Aloe humilis* (L.) Mill.
Aloe humilis (L.) Mill. var. *suberecta* (Aiton) Baker. . . . . *Aloe humilis* (L.) Mill.
Aloe humilis (L.) Mill. var. *subtuberculata* (Haw.) Baker . . . . *Aloe humilis* (L.) Mill.
Aloe humilis Ker Gawl. (nom. illegit.) . . . . *Aloe humilis* (L.) Mill.

Aloe immaculata Pillans . . . .*Aloe affinis* A.Berger
Aloe incurva (Haw.) Haw. . . . . *Aloe humilis* (L.) Mill.
Aloe indica Royle . . . . *Aloe vera* (L.) Burm.f.
Aloe intermedia (H.Perrier) Reynolds (nom. illegit.). . . . . *Aloe newtonii* J.-B.Castillon
Aloe itremensis Reynolds. . . . .*Aloe ibitiensis* H.Perrier

Aloe jex-blakeae Christian . . . . *Aloe ruspoliana* Baker
Aloe johnstonii Baker . . . . *Aloe myriacantha* (Haw.) Schult. & Schult.f.
Aloe juttae Dinter . . . . *Aloe microstigma* Salm-Dyck

Aloe keithii Reynolds. . . . .*Aloe parvibracteata* Schönland
Aloe kirkii Baker . . . . *Aloe leptosiphon* A.Berger
Aloe koenenii Lavranos & Kerstin Koch . . . *Aloe porphyrostachys* Lavranos & Collen. subsp. *koenenii* (Lavranos & Kerstin Koch) Lodé
Aloe krapohliana Marloth var. *dumoulinii* Lavranos. . . . . *Aloe krapohliana* Marloth
Aloe kraussii Baker var. *minor* Baker. . . . . *Aloe albida* (Stapf) Reynolds
Aloe kraussii Schönland (nom. illegit.). . . . . *Aloe albida* (Stapf) Reynolds

Aloe labiaflava Groenew. . . . . *Aloe davyana* Schönland
Aloe lanuriensis De Wild. . . . . *Aloe macrocarpa* Tod. subsp. *wollastonii* (Rendle) Wabuyele
Aloe lanzae Tod. . . . . *Aloe vera* (L.) Burm.f.

Aloe lastii Baker . . . . *Aloe brachystachys* Baker
Aloe lateritia Engl. var. *kitaliensis* (Reynolds) Reynolds. . . . . . *Aloe macrocarpa* Tod. subsp. *wollastonii* (Rendle) Wabuyele
Aloe latifolia (Haw.) Haw. . . . . *Aloe maculata* All.
Aloe laxiflora N.E.Br.. . . . . *Aloe gracilis* Haw.
Aloe laxissima Reynolds . . . . *Aloe transvaalensis* Kuntze
Aloe leachii Reynolds . . . . *Aloe sobolifera* (S.Carter) Wabuyele
Aloe leptophylla N.E.Br. ex Baker . . . . *Aloe maculata* All.
Aloe leptophylla N.E.Br. ex Baker var. *stenophylla* Baker. . . . . *Aloe maculata* All.
Aloe lineata (Aiton) Haw. var. *glaucescens* Haw. . . . . *Aloe lineata* (Aiton) Haw. var. *lineata*
Aloe lineata (Aiton) Haw. var. *viridis* Haw. . . . . *Aloe lineata* (Aiton) Haw. var. *lineata*
Aloe lingua Thunb. . . . . *Aloe plicatilis* (L.) Mill.
Aloe linguaeformis L.f. (nom. illegit.) . . . . *Aloe plicatilis* (L.) Mill.
Aloe longiaristata Schult. & Schult.f.. . . . . *Aloe aristata* Haw.
Aloe lugardiana Baker . . . . *Aloe zebrina* Baker
Aloe lusitanica Groenew. . . . .*Aloe parvibracteata* Schönland

Aloe macilenta (Baker) G.Nicholson . . . . *Aloe humilis* (L.) Mill.
Aloe macowanii Baker . . . . *Aloe striatula* Haw. var. *striatula*
Aloe macracantha Baker . . . . *Aloe maculata* All.
Aloe macrocarpa Tod. var. *major* A.Berger . . . . . . *Aloe macrocarpa* Tod. subsp. *macrocarpa*
Aloe maculata All. var. *ficksburgensis* (Reynolds) Dandy . . . . *Aloe maculata* All.
Aloe maculata Forrsk. (nom. illegit.) . . . . *Aloe vera* (L.) Burm.f.
Aloe maculosa Lam. . . . . *Aloe maculata* All.
Aloe magnidentata I.Verd. & Christian. . . . .*Aloe megalacantha* Baker subsp. *megalacantha*
Aloe marginalis DC. (nom. superfl.) . . . . *Aloe purpurea* Lam.
Aloe marginata (Aiton) Willd. (nom. illegit.) . . . . *Aloe purpurea* Lam.
Aloe marlothii A.Berger var. *bicolor* Reynolds. . . . . *Aloe marlothii* A.Berger subsp. *marlothii*
Aloe marsabitensis I.Verd. & Christian. . . . .*Aloe secundiflora* Engl. var. *secundiflora*
Aloe marshallii J.M.Wood & M.S.Evans. . . . .*Aloe kniphofioides* Baker
Aloe melanacantha A.Berger var. *erinacea* (D.S.Hardy) G.D.Rowley . . . .*Aloe erinacea* D.S.Hardy
Aloe melsetterensis Christian . . . . *Aloe swynnertonii* Rendle
Aloe meruana Lavranos . . . .*Aloe chrysostachys* Lavranos & L.E.Newton
Aloe microstigma Salm-Dyck subsp. *framesii* (L.Bolus) Glen & D.S.Hardy. . . . *Aloe framesii* L.Bolus
Aloe minima Baker var. *blyderivierensis* (Groenew.) Reynolds . . . . *Aloe minima* Baker
Aloe minima J.M.Wood (nom. illegit.). . . . . *Aloe saundersiae* (Reynolds) Reynolds
Aloe mitraeformis DC. (nom. illegit.). . . . . *Aloe perfoliata* L.
Aloe mitraeformis Salm-Dyck . . . . *Aloe perfoliata* L.

Aloe mitraeformis Willd. (nom. illegit.) . . . . *Aloe perfoliata* L.
Aloe mitriformis Mill. . . . . *Aloe perfoliata* L.
Aloe mitriformis Mill. subsp. *comptonii* (Reynolds) Zonn. . . . . *Aloe comptonii* Reynolds
Aloe mitriformis Mill. subsp. *distans* (Haw.) Zonn. . . . .*Aloe distans* Haw.
Aloe mitriformis Mill. var. *albispina* (Haw.) A.Berger. . . . . *Aloe perfoliata* L.
Aloe mitriformis Mill. var. *angustior* Lam. . . . .*Aloe distans* Haw.
Aloe mitriformis Mill. var. *brevifolia* (Aiton) W.T.Aiton . . . .*Aloe distans* Haw.
Aloe mitriformis Mill. var. *commelinii* (Willd.) Baker. . . . . *Aloe perfoliata* L.
Aloe mitriformis Mill. var. *elatior* Haw. . . . . *Aloe perfoliata* L.
Aloe mitriformis Mill. var. *flavispina* (Haw.) Baker . . . . *Aloe perfoliata* L.
Aloe mitriformis Mill. var. *humilior* Haw.. . . . . *Aloe perfoliata* L.
Aloe mitriformis Mill. var. *humilior* Willd. . . . .*Aloe distans* Haw.
Aloe mitriformis Mill. var. *pachyphylla* Baker. . . . . *Aloe perfoliata* L.
Aloe mitriformis Mill. var. *spinosior* Haw. . . . .*Aloe ×nobilis* Haw.
Aloe mitriformis Mill. var. *spinulosa* (Salm-Dyck) Baker . . . . *Aloe perfoliata* L.
Aloe mitriformis Mill. var. *xanthacantha* (Willd.) Baker . . . . *Aloe perfoliata* L.
Aloe mketiensis Christian . . . . *Aloe nuttii* Baker
Aloe montana Schinz . . . . *Aloe dichotoma* Masson
Aloe morogoroensis Christian . . . .*Aloe bussei* A.Berger
Aloe muirii Marloth. . . . . *Aloe lineata* (Aiton) Haw. var. *muirii* (Marloth) Reynolds
Aloe muricata Haw. . . . .*Aloe ferox* Mill.
Aloe muricata Schult. (nom. illegit.) . . . . *Aloe glauca* Mill.
Aloe mwanzana Christian. . . . . *Aloe macrosiphon* Baker
Aloe myriacantha (Haw.) Schult. & Schult.f. var. *minor* (Baker) A.Berger. . . . . . *Aloe albida* (Stapf) Reynolds

Aloe natalensis J.M.Wood & M.S.Evans . . . . . . *Aloe arborescens* Mill. subsp. *arborescens*
Aloe ngobitensis Reynolds . . . . *Aloe nyeriensis* Christian
Aloe nitens Baker (nom. illegit.) . . . . *Aloe rupestris* Baker
Aloe nobilis Baker (nom. illegit.). . . . . *Aloe stans* A.Berger (exact application unknown)
Aloe nobilis Baker var. *densifolia* Baker. . . .*Aloe brownii* Baker (exact application unknown)
Aloe nyeriensis Christian subsp. *kedongensis* (Reynolds) S.Carter . . . . . . *Aloe kedongensis* Reynolds

Aloe obscura A.Berger ex Schönland (nom. illegit.) . . . . *Aloe ×runcinata* A.Berger
Aloe obscura Mill.. . . . . *Aloe maculata* All.
Aloe oligospila Baker. . . . .*Aloe percrassa* Tod.
Aloe orpeniae Schönland . . . .*Aloe hereroensis* Engl. var. *hereroensis*
Aloe otallensis Baker var. *elongata* A.Berger. . . . . . . *Aloe rugosifolia* M.G.Gilbert & Sebsebe

Aloe pallescens Haw. . . . . *Aloe serrulata* (Aiton) Baker (exact application unknown)
Aloe pallidiflora A.Berger. . . . . *Aloe greatheadii* Schönland
Aloe paniculata Jacq. . . . . *Aloe striata* Haw.
Aloe paradoxa A.Berger . . . . *Aloe ×heteracantha* Baker
Aloe parvibracteata Schönland var. *zuluensis* (Reynolds) Reynolds. . . . *Aloe parvibracteata* Schönland
Aloe parvicapsula Lavranos & Collen. . . . .*Aloe woodii* Lavranos & Collen.
Aloe parvicoma Lavranos & Collen. . . . . *Aloe rivierei* Lavranos & L.E.Newton
Aloe parviflora Baker. . . . . *Aloe minima* Baker
Aloe parvispina Schönland. . . . . *Aloe perfoliata* L.
Aloe ×paxii A.Terracc.. . . . . *Aloe ×schimperi* Tod.
Aloe peacockii (Baker) A.Berger . . . . *Aloe elegans* Tod.
Aloe pentagona Salm-Dyck . . . .*Aloe quinquangularis* Schult.f. (exact application unknown)
Aloe percrassa A.Berger (nom. illegit.) var. *saganeitiana* A.Berger . . . . . . *Aloe elegans* Tod.
Aloe percrassa Schweinf. var. *albo-picta* Schweinf.. . . . . *Aloe trichosantha* A.Berger subsp. *trichosantha*
Aloe percrassa Schweinf. var. *menachensis* Schweinf.. . . . . *Aloe ×menachensis* (Schweinf.) Blatt.
Aloe perfoliata L. var. [α] *arborescens* (Mill.) Aiton . . . .*Aloe arborescens* Mill. subsp. *arborescens*
Aloe perfoliata L. var. [β] *africana* (Mill.) Aiton. . . . . *Aloe africana* Mill.
Aloe perfoliata L. var. [γ] *barbadensis* (Mill.) Aiton . . . . *Aloe vera* (L.) Burm.f.
Aloe perfoliata L. var. *brevifolia* Aiton . . . .*Aloe distans* Haw.
Aloe perfoliata L. var. [θ] *ferox* (Mill.) Aiton . . . .*Aloe ferox* Mill.
Aloe perfoliata L. var. [ζ] *glauca* (Mill.) Aiton. *Aloe glauca* Mill.
Aloe perfoliata L. var. [o] *humilis* . . . . *Aloe humilis* (L.) Mill.
Aloe perfoliata L. var. [η] *lineata* Aiton . . . . *Aloe lineata* (Aiton) Haw. var. *lineata*
Aloe perfoliata L. var. *mitriformis* (Mill.) Aiton . . . . *Aloe perfoliata* L.
Aloe perfoliata L. var. *obscura* (Mill.) Aiton . . . . *Aloe maculata* All.
Aloe perfoliata L. var. *purpurascens* Aiton . . . .*Aloe succotrina* Weston
Aloe perfoliata L. var. [τ] *saponaria* Aiton . . . . *Aloe maculata* All.
Aloe perfoliata L. var. *serrulata* Aiton . . . . . . *Aloe serrulata* (Aiton) Baker (exact application unknown)
Aloe perfoliata L. var. [μ] *suberecta* Aiton . . . . *Aloe humilis* (L.) Mill.
Aloe perfoliata L. var. *succotrina*(Lam.) Aiton . . . .*Aloe succotrina* Weston
Aloe perfoliata L. var. [λ] *vera* Willd.. . . . . *Aloe vera* (L.) Burm.f.
Aloe perfoliata L. var. [π] *vera* L. . . . . *Aloe vera* (L.) Burm.f.
Aloe perfoliata L. var. α L. . . . . *Aloe commixta* A.Berger
Aloe perfoliata L. var. β L. . . . . *Aloe africana* Mill.
Aloe perfoliata L. var. γ L. . . . .*Aloe ferox* Mill.

Aloe perfoliata L. var. δ L. . . . . *Aloe brevifolia* Mill. var. *brevifolia*

Aloe perfoliata L. var. ε L. . . . . *Aloe ferox* Mill.

Aloe perfoliata L. var. ζ L. . . . . *Aloe brevifolia* Mill. var. *brevifolia*

Aloe perfoliata L. var. ζ Willd. . . . . *Aloe ferox* Mill.

Aloe perfoliata L. var. η L. . . . . *Aloe arborescens* Mill. subsp. *arborescens*

Aloe perfoliata L. var. θ L. . . . . *Aloe maculata* All.

Aloe perfoliata L. var. κ L. . . . . *Aloe glauca* Mill.

Aloe perfoliata L. var. κ Willd. . . . . *Aloe perfoliata* L.

Aloe perfoliata L. var. λ L. . . . . *Aloe maculata* All.

Aloe perfoliata L. var. *μ* L. . . . . *Aloe maculata* All.

Aloe perfoliata L. var. ν L. . . . . *Aloe perfoliata* L.

Aloe perfoliata L. var. ξ L. . . . . *Aloe succotrina* Weston

Aloe perfoliata L. var. ξ Willd. . . . . *Aloe perfoliata* L.

Aloe perfoliata Thunb.. . . . . *Aloe ferox* Mill.

Aloe picta Thunb. . . . . *Aloe maculata* All.

Aloe picta Thunb. var. *major* Willd.. . . . . *Aloe maculata* All.

Aloe platylepis Baker . . . . *Aloe hexapetala* Salm-Dyck (exact application unknown)

Aloe platyphylla Baker. . . . . *Aloe zebrina* Baker

Aloe plicatilis (L.) Mill. var. *major* Salm-Dyck. . . . . *Aloe plicatilis* (L.) Mill.

Aloe pluridens Haw. var. *beckeri* Schönland. . . . . *Aloe pluridens* Haw.

Aloe pole-evansii Christian. . . . . *Aloe dawei* A.Berger

Aloe pongolensis Reynolds. . . . . *Aloe parvibracteata* Schönland

Aloe pongolensis Reynolds var. *zuluensis* Reynolds. . . . . *Aloe parvibracteata* Schönland

Aloe postgenita Schult. & Schult.f. . . . . *Aloe brevifolia* Mill. var. *brevifolia*

Aloe prolifera Haw.. . . . . *Aloe brevifolia* Mill. var. *brevifolia*

Aloe prolifera Haw. var. *major* Salm-Dyck . . . . *Aloe brevifolia* Mill. var. *brevifolia*

Aloe propagulifera (Rauh & Razaf.) L.E.Newton & G.D.Rowley. . . . . *Aloe prostrata* (H.Perrier) L.E.Newton & G.D.Rowley

Aloe prostrata (H.Perrier) L.E.Newton & G.D.Rowley subsp. *pallida* Rauh. . .*Aloe sakarahensis* Lavranos & M.Teissier subsp. *pallida* (Rauh) Lavranos & M.Teissier

Aloe pseudoafricana Salm-Dyck . . . . *Aloe africana* Mill.

Aloe pseudo-ferox Salm-Dyck. . . . . *Aloe ferox* Mill.

Aloe pulchra Lavranos (nom. illegit.). . . . . *Aloe bella* G.D.Rowley

Aloe punctata Haw. . . . . *Aloe variegata* L.

Aloe purpurascens (Aiton) Haw. . . . . *Aloe succotrina* Weston

Aloe pycnantha MacOwen . . . . *Aloe rupestris* Baker

Aloe ramosa Haw.. . . . . *Aloe dichotoma* Masson

Aloe recurvifolia Groenew.. . . . . *Aloe alooides* (Bolus) Druten

Aloe reflexa van Marum ex Steud. . . . . *Aloe distans* Haw.

Aloe rhodacantha DC. . . . . *Aloe glauca* Mill.
Aloe rhodocincta Baker . . . . *Aloe striata* Haw.
Aloe richtersveldensis Venter & Beukes . . . . *Aloe meyeri* van Jaarsv.
Aloe rigens Reynolds & P.R.O.Bally var. *glabrescens* Reynolds & P.R.O.Bally. . . *Aloe glabrescens* (Reynolds & P.R.O.Bally) S.Carter & Brandham
Aloe rossi Tod. . . . . *Aloe deltoideodonta* Baker var. *deltoideodonta*
Aloe rubescens DC.. . . . . *Aloe vera* (L.) Burm.f.
Aloe rubrolutea Schinz. . . . . *Aloe littoralis* Baker
Aloe ruffingiana Rauh & Petignat . . . . . *Aloe deltoideodonta* Baker var. *ruffingiana* (Rauh & Petignat) J.-B.Castillon & J.-P.Castillon
Aloe rufocincta Haw.. . . . . *Aloe purpurea* Lam.
Aloe ruspoliana Baker var. *draceniformis* A.Berger . . . . *Aloe retrospiciens* Reynolds

Aloe sabila Karw. ex Steud. . . . . *Aloe vera* (L.) Burm.f.
Aloe sakoankenke J.-B.Castillon . . . . .*Aloe massawana* Reynolds subsp. *sakoankenke* (J.-B. Castillon) J.-B.Castillon
Aloe ×*salm-dyckiana* Schult. & Schult.f. . . . . *Aloe* ×*principis* (Haw.) Stearn
Aloe saponaria (Aiton) Haw.. . . . . *Aloe maculata* All.
Aloe saponaria (Aiton) Haw. var. *brachyphylla* Baker. . . . . *Aloe maculata* All.
Aloe saponaria (Aiton) Haw. var. *ficksburgensis* Reynolds . . . . *Aloe maculata* All.
Aloe saponaria (Aiton) Haw. var. *latifolia* Haw.. . . . . *Aloe maculata* All.
Aloe saponaria (Aiton) Haw. var. *obscura* (Mill.) Haw.. . . . . *Aloe maculata* All.
Aloe saronarae Lavranos & T.A.McCoy. . . . . *Aloe ibitiensis* H.Perrier
Aloe schimperi G.Karst. & Schenck (nom. illegit.). . . . .*Aloe percrassa* Tod.
Aloe schimperi Schweinf. (nom. illegit.). . . . .*Aloe percrassa* Tod.
Aloe schinzii Baker . . . . *Aloe littoralis* Baker
Aloe schlechteri Schönland . . . . *Aloe claviflora* Burch.
Aloe schliebenii Lavranos . . . . *Aloe brachystachys* Baker
Aloe schmidtiana Regel . . . . *Aloe cooperi* Baker subsp. *cooperi*
Aloe schweinfurthii Baker var. *labworana* Reynolds. . . . *Aloe labworana* (Reynolds) S.Carter
Aloe schweinfurthii Hook.f. (nom. illegit.) . . . . *Aloe elegans* Tod.
Aloe secundiflora Engl. var. *sobolifera* S.Carter . . . . . . *Aloe sobolifera* (S.Carter) Wabuyele
Aloe sempervivoides H.Perrier . . . . *Aloe parvula* A.Berger
Aloe serra DC. . . . . *Aloe brevifolia* Mill. var. *depressa* (Haw.) Baker
Aloe sessiliflora Pole-Evans . . . . *Aloe spicata* L.f.
Aloe sigmoidea Baker (possibly a hybrid). . . . . . *Aloe arborescens* Mill. subsp. *arborescens*
Aloe sinuata Thunb. . . . . *Aloe succotrina* Weston
Aloe soccotorina Schult. & Schult.f. . . . . *Aloe succotrina* Weston
Aloe soccotrina Garsault . . . . *Aloe succotrina* Weston
Aloe socotrina DC. . . . . *Aloe succotrina* Weston

Aloe socotrina DC. var. [β] *purpurascens* (Aiton) Ker Gawl. . . . . *Aloe succotrina* Weston

Aloe solaiana Christian . . . . *Aloe lateritia* Engl. var. *graminicola* (Reynolds) S.Carter

Aloe spicata Baker (nom. illegit.) . . . . *Aloe camperi* Schweinf.

Aloe spinulosa Salm-Dyck . . . . *Aloe perfoliata* L.

Aloe steffaniana Rauh. . . *Aloe versicolor* Guillaumin var. *steffaniana* (Rauh) J.-B.Castillon & J.-P.Castillon

Aloe stefaninii Chiov. . . . . *Aloe ruspoliana* Baker

Aloe striata Haw. subsp. *karasbergensis* (Pillans) Glen & D.S.Hardy. . . *Aloe karasbergensis* Pillans

Aloe striata Haw. subsp. *komaggasensis* (Kritz. & van Jaarsv.) Glen & D.S.Hardy. . . . . *Aloe komaggasensis* Kritz. & van Jaarsv.

Aloe striata Haw. var. *oligospila* Baker . . . . *Aloe striata* Haw.

Aloe striata Haw. var. *rhodocincta* (Baker) Trel. . . . . *Aloe striata* Haw.

Aloe stuhlmannii Baker . . . . *Aloe volkensii* Engl. subsp. *volkensii*

Aloe suberecta (Aiton) Haw.. . . . . *Aloe humilis* (L.) Mill.

Aloe suberecta (Aiton) Haw. var. *acuminata* Haw. . . . . *Aloe humilis* (L.) Mill.

Aloe suberecta (Aiton) Haw. var. *semiguttata* Haw. . . . . *Aloe humilis* (L.) Mill.

Aloe subferox Spreng. . . . . *Aloe ferox* Mill.

Aloe subinermis Lem. . . . . *Aloe striatula* Haw. var. *striatula*

Aloe subtuberculata Haw. . . . . *Aloe humilis* (L.) Mill.

Aloe succotrina Lam. . . . . *Aloe succotrina* Weston

Aloe succotrina Lam. var. *saxigena* A.Berger . . . . *Aloe succotrina* Weston

Aloe supralaevis Haw. . . . . *Aloe ferox* Mill.

Aloe supralaevis Haw. var. *erythrocarpa* Baker . . . . *Aloe ferox* Mill.

Aloe supralaevis Haw. var. *hanburyi* Baker . . . . . . *Aloe marlothii* A.Berger subsp. *marlothii*

Aloe tenuior Haw. var. *decidua* Reynolds. . . . . *Aloe tenuior* Haw.

Aloe tenuior Haw. var. *densiflora* Reynolds . . . . *Aloe tenuior* Haw.

Aloe tenuior Haw. var. *glaucescens* Zahlbr. . . . . *Aloe tenuior* Haw.

Aloe tenuior Haw. var. *rubriflora* Reynolds. . . . . *Aloe tenuior* Haw.

Aloe tenuior Haw. var. *viridifolia* van Jaarsv. *Aloe tenuior* Haw.

Aloe termetophyla De Wild. . . . . *Aloe greatheadii* Schönland

Aloe tidmarshii (Schönland) F.S.Mull. ex R.A.Dyer . . . . . . *Aloe ciliaris* Haw. var. *tidmarshii* Schönland

Aloe torrei I.Verd. & Christian var. *wildii* Reynolds. . . . . *Aloe wildii* (Reynolds) Reynolds

Aloe transvaalensis Kuntze var. *stenacantha* Groenew. . . . . *Aloe transvaalensis* Kuntze

Aloe trichosantha A.Berger var. *menachensis* (Schweinf.) A.Berger. . . . *Aloe* ×*menachensis* (Schweinf.) Blatt.

Aloe trichotoma Colla . . . . *Aloe maculata* All.

Aloe tricolor Baker (nom. illegit.) . . . . *Aloe maculata* All.

Aloe tripetala Medik.. . . . . *Aloe plicatilis* (L.) Mill.

Aloe trothae A.Berger . . . . *Aloe bulbicaulis* Christian

Aloe tuberculata Haw.. . . . . *Aloe humilis* (L.) Mill.

Aloe tugenensis L.E.Newton & Lavranos . . . . *Aloe archeri* Lavranos subsp. *tugenensis* (L.E.Newton & Lavranos) Wabuyele

Aloe tweediae Christian . . . . *Aloe secundiflora* Engl. var. *tweediae* (Christian) Wabuyele

Aloe umbellata DC.. . . . . *Aloe maculata* All.

Aloe vaotsohy Decorse & Poiss. . . . . *Aloe divaricata* A.Berger var. *divaricata*

Aloe vaotsohy Decorse & Poiss. var. *rosea* Decary . . . . . . *Aloe divaricata* A.Berger var. *rosea* (Decary) Reynolds

Aloe variegata Forssk (nom. illegit.) . . . . *Aloe vera* (L.) Burm.f.

Aloe variegata L. var. *haworthii* A.Berger. . . . . *Aloe variegata* L.

Aloe venusta Reynolds (nom. illegit.). . . . .*Aloe bicomitum* L.C.Leach

Aloe vera L. var. *aethiopica* Schweinf.. . . . . *Aloe elegans* Tod.

Aloe vera L. var. *angustifolia* Schweinf. *Aloe officinalis* Forssk. var. *angustifolia* (Schweinf.) Lavranos

Aloe vera L. var. *chinensis* (Steud. ex Baker) Baker. . . . . *Aloe vera* (L.) Burm.f.

Aloe vera L. var. *lanzae* (Tod.) Baker. . . . . *Aloe vera* (L.) Burm.f.

Aloe vera L. var. *officinalis* (Forssk.) Baker . . . . *Aloe officinalis* Forssk. var. *officinalis*

Aloe vera L. var. *puberula* Schweinf. . . . . *Aloe* ×*puberula* (Schweinf.) A.Berger

Aloe vera Mill. (nom. illegit.) . . . .*Aloe succotrina* Weston

Aloe verrucosospinosa All. . . . . *Aloe humilis* (L.) Mill.

Aloe virens Haw. var. *macilenta* Baker f. . . . . *Aloe virens* Haw. (exact application unknown)

Aloe vulgaris Lam. . . . . *Aloe vera* (L.) Burm.f.

Aloe vulgaris Lam. var. *abyssinica* DC. . . . .*Aloe abyssinica* Lam. (exact application unknown)

Aloe wollastonii Rendle . . . . . . *Aloe macrocarpa* Tod. subsp. *wollastonii* (Rendle) Wabuyele

Aloe woolliana Pole-Evans . . . . *Aloe chortolirioides* A.Berger var. *woolliana* (Pole-Evans) Glen & D.S.Hardy

Aloe xanthacantha Salm-Dyck (nom. illegit.) . . . . *Aloe perfoliata* L.

Aloe xanthacantha Willd.. . . . . *Aloe perfoliata* L.

Aloe zanzibarica Milne-Redh.. . . . . *Aloe squarrosa* Baker ex Balf.f.

Aloe zeyheri Baker (nom. illegit.) . . . . *Aloe barberae* Dyer

Aloe zombitsiensis Rauh & M.Teissier . .*Aloe prostrata* (H.Perrier) L.E.Newton & G.D.Rowley

Bowiea africana Haw. . . . . *Aloe bowiea* Schult. & Schult.f.
Bowiea myriacantha Haw. . . . . *Aloe myriacantha* (Haw.) Schult. & Schult.f.

Catevala arborescens (Mill.) Medik. . . . . *Aloe arborescens* Mill. subsp. *arborescens*
Catevala humilis (L.) Medik. . . . . *Aloe humilis* (L.) Mill.
Chamaealoe africana (Haw.) A.Berger. . . . . *Aloe bowiea* Schult. & Schult.f.

Gasteria antandroi Decary . . . . *Aloe antandroi* (Decary) H.Perrier subsp. *antandroi*
Guillauminia albiflora (Guillaumin) A.Bertrand. . . . . *Aloe albiflora* Guillaumin
Guillauminia bakeri (Scott-Elliot) P.V.Heath . . . . *Aloe bakeri* Scott-Elliot
Guillauminia bellatula (Reynolds) P.V.Heath . . . . *Aloe bellatula* Reynolds
Guillauminia calcairophila (Reynolds) P.V.Heath. . . . . *Aloe calcairophila* Reynolds
Guillauminia descoingsii (Reynolds) P.V.Heath . . . . *Aloe descoingsii* Reynolds subsp. *descoingsii*
Guillauminia rauhii (Reynolds) P.V.Heath. . . . . *Aloe rauhii* Reynolds

Kumara disticha Medik.. . . . . *Aloe plicatilis* (L.) Mill.

Leemea boiteaui (Guillaumin) P.V.Heath . . . . *Aloe boiteaui* Guillaumin
Leemea haworthioides (Baker) P.V.Heath . . . . . *Aloe haworthioides* Baker var. *haworthioides*
Leemea parvula (A.Berger) P.V.Heath . . . . *Aloe parvula* A.Berger
Leptaloe albida Stapf . . . . *Aloe albida* (Stapf) Reynolds
Leptaloe blyderivierensis Groenew. . . . . *Aloe minima* Baker
Leptaloe caricina (A.Berger) Stapf . . . . *Aloe myriacantha* (Haw.) Schult. & Schult.f.
Leptaloe graminifolia (A.Berger) Stapf. . . . . . . *Aloe myriacantha* (Haw.) Schult. & Schult.f.
Leptaloe johnstonii (Baker) Christian. . . . . *Aloe myriacantha* (Haw.) Schult. & Schult.f.
Leptaloe minima (Baker) Stapf . . . . *Aloe minima* Baker
Leptaloe myriacantha (Haw.) Stapf. . . . . *Aloe myriacantha* (Haw.) Schult. & Schult.f.
Leptaloe parviflora (Baker) Stapf . . . . *Aloe minima* Baker
Leptaloe saundersiae Reynolds. . . . . *Aloe saundersiae* (Reynolds) Reynolds
Lomatophyllum aldabrense Marais . . . .*Aloe aldabrensis* (Marais) L.E.Newton & G.D.Rowley
Lomatophyllum aloiflorum (Ker Gawl.) G.Nicholson . . . . *Aloe purpurea* Lam.
Lomatophyllum anivoranoense Rauh & Hebding. . . . *Aloe anivoranoensis* (Rauh & Hebding) L.E.Newton & G.D.Rowley
Lomatophyllum antsingyense Léandri . . . .*Aloe antsingyensis* (Léandri) L.E.Newton & G.D.Rowley
Lomatophyllum belavenokense Rauh & Gerold. . . . . . *Aloe belavenokensis* (Rauh & Gerold) L.E.Newton & G.D.Rowley
Lomatophyllum borbonicum Willd. (nom. superfl.) . . . . *Aloe purpurea* Lam.
Lomatophyllum citreum Guillaumin . . . *Aloe citrea* (Guillaumin) L.E.Newton & G.D.Rowley

Lomatophyllum lomatophylloides (Balf.f) Marais. . . . . *Aloe lomatophylloides* Balf.f.
Lomatophyllum macrum (Haw.) Salm-Dyck ex Schult. & Schult.f.. . . . . . . *Aloe macra* Haw.
Lomatophyllum namorokaense Rauh. . . . . *Aloe namorokaensis* (Rauh) L.E.Newton & G.D.Rowley
Lomatophyllum occidentale H.Perrier . . . .*Aloe occidentalis* (H.Perrier) L.E.Newton & G.D.Rowley
Lomatophyllum oligophyllum (Baker) H.Perrier. . . . .*Aloe oligophylla* Baker
Lomatophyllum orientale H.Perrier . . . . *Aloe orientalis* (H.Perrier) L.E.Newton & G.D.Rowley
Lomatophyllum pembanum (L.E.Newton) Rauh . . . . *Aloe pembana* L.E.Newton
Lomatophyllum peyrierasii (Cremers) Rauh . . . . *Aloe peyrierasii* Cremers
Lomatophyllum propaguliferum Rauh & Razaf. . . . .*Aloe propagulifera* (Rauh & Razaf.) L.E.Newton & G.D.Rowley
Lomatophyllum prostratum H.Perrier. . . . . *Aloe prostrata* (H.Perrier) L.E.Newton & G.D.Rowley
Lomatophyllum purpureaum (Lam.) T.Durand & Schinz. . . . . *Aloe purpurea* Lam.
Lomatophyllum roseum H.Perrier . . . . *Aloe rosea* (H.Perrier) L.E.Newton & G.D.Rowley
Lomatophyllum rufocinctum (Haw.) Salm-Dyck ex Schult. & Schult.f. . . . . .*Aloe purpurea* Lam
Lomatophyllum sociale H.Perrier . . . . . .*Aloe socialis* (H.Perrier) L.E.Newton & G.D.Rowley
Lomatophyllum tormentorii Marais. . . . *Aloe tormentorii* (Marais) L.E.Newton & G.D.Rowley
Lomatophyllum viviparum H.Perrier . . . .*Aloe schilliana* L.E.Newton & G.D.Rowley

Notosceptrum alooides (Bolus) Benth. . . . . *Aloe alooides* (Bolus) Druten

Pachidendron africanum (Mill.) Haw. . . . . *Aloe africana* Mill.
Pachidendron africanum (Mill.) Haw. var. *angustum* Haw. . . . . *Aloe africana* Mill.
Pachidendron africanum (Mill.) Haw. var. *latum* Haw. . . . . *Aloe africana* Mill.
Pachidendron angustifolium (Haw.) Haw. . . . . *Aloe africana* Mill.
Pachidendron ferox (Mill.) Haw. . . . .*Aloe ferox* Mill.
Pachidendron principis Haw. . . . . *Aloe* ×*principis* (Haw.) Stearn
Pachidendron pseudo-ferox (Salm-Dyck) Haw.. . . . .*Aloe ferox* Mill.
Pachidendron supralaeve (Haw.) Haw. . . . .*Aloe ferox* Mill.
Phylloma aloiflorum Ker Gawl. (nom. illegit.) . . . . *Aloe purpurea* Lam.
Phylloma macrum (Haw.) Sweet. . . . . *Aloe macra* Haw.
Phylloma rufocinctum (Haw.) Sweet . . . .*Aloe purpurea* Lam

Rhipidodendron dichotomum (Masson) Willd. . . . . *Aloe dichotoma* Masson
Rhipidodendron distichum (Medik.) Willd. . . . . *Aloe plicatilis* (L.) Mill.
Rhipidodendron plicatile (L.) Haw.. . . . . *Aloe plicatilis* (L.) Mill.

Urginea alooides Bolus . . . . *Aloe alooides* (Bolus) Druten

Common names

Common name	Accepted name
//noru	*Aloe zebrina* Baker
//nuru	*Aloe zebrina* Baker
/ganya	*Aloe zebrina* Baker
/gikwe	*Aloe zebrina* Baker
| |ganja	*Aloe zebrina* Baker
| |garas	*Aloe dichotoma* Masson
| |gores	*Aloe asperifolia* A.Berger
aalewee	*Aloe* L.
aalwee	*Aloe* L.
aalwyn	*Aloe* L.
aanteel-aalwyn	*Aloe brevifolia* Mill. var. *brevifolia*
aanteelaalwyn	*Aloe claviflora* Burch. *Aloe grandidentata* Salm-Dyck
aboes	*Aloe vera* (L.) Burm.f.
açevar	*Aloe succotrina* Weston
achuka	*Aloe secundiflora* Engl. var. *secundiflora* *Aloe secundiflora* Engl. var. *tweediae* (Christian) Wabuyele
acibar	*Aloe vera* (L.) Burm.f.
acíbar	*Aloe vera* (L.) Burm.f.
acibara	*Aloe ferox* Mill.
afgeronde-aalwyn	*Aloe gariepensis* Pillans
agave	*Aloe vera* (L.) Burm.f.
aloe a candelabro	*Aloe arborescens* Mill. var. *arborescens*
aloé arborescente	*Aloe arborescens* Mill. var. *arborescens*
aloë boom	*Aloe khamiesensis* Pillans
aloé candelabro	*Aloe arborescens* Mill. var. *arborescens*
áloe de los Barbados	*Aloe vera* (L.) Burm.f.
áloe del cabo	*Aloe ferox* Mill.
aloe delle Barbados	*Aloe vera* (L.) Burm.f.
aloe di Curacao	*Aloe vera* (L.) Burm.f.
aloé dos Libombos	*Aloe spicata* L.f.
áloe estriada	*Aloe succotrina* Weston
aloe mediterranea	*Aloe vera* (L.) Burm.f.
aloe of the shore	*Aloe littoralis* Baker
áloe socotrina	*Aloe succotrina* Weston
aloe vera	*Aloe vera* (L.) Burm.f.

Common name	Scientific name
aloé vera	*Aloe vera* (L.) Burm.f.
aloe with hair-like tufts	*Aloe comosa* Marloth & A.Berger
aloé	*Aloe arborescens* Mill. var. *arborescens*
	Aloe vera (L.) Burm.f.
aloe	*Aloe* L.
	Aloe vera (L.) Burm.f.
aloë	*Aloe* L.
	Aloe vera (L.) Burm.f.
áloe	*Aloe maculata* All.
	Aloe vera (L.) Burm.f.
aloeboom	*Aloe khamiesensis* Pillans
aloé-candelabro	*Aloe arborescens* Mill. var. *arborescens*
aloe-capim	*Aloe cooperi* Baker subsp. *pulchra* Glen & D.S.Hardy
aloé-de-Barbados	*Aloe vera* (L.) Burm.f.
aloé-dos-Barbados	*Aloe vera* (L.) Burm.f.
aloès amer	*Aloe vera* (L.) Burm.f.
aloès arborescent	*Aloe arborescens* Mill. var. *arborescens*
aloés de Barbados	*Aloe vera* (L.) Burm.f.
aloès du Cap	*Aloe ferox* Mill.
aloes mydlnicowaty	*Aloe maculata* All.
aloes pregowany	*Aloe zebrina* Baker
aloès tacheté	*Aloe zebrina* Baker
aloès vulgaire	*Aloe vera* (L.) Burm.f.
aloès zébré	*Aloe zebrina* Baker
aloes zwyczajny	*Aloe vera* (L.) Burm.f.
aloés	*Aloe arborescens* Mill. var. *arborescens*
	Aloe vera (L.) Burm.f.
aloes	*Aloe* L.
	Aloe vera (L.) Burm.f.
	Aloe zebrina Baker
aloès	*Aloe vera* (L.) Burm.f.
aloja	*Aloe vera* (L.) Burm.f.
amahlala	*Aloe maculata* All.
amaposo	*Aloe arborescens* Mill. var. *arborescens*
ananas marron	*Aloe lomatophylloides* Balf.f.
ananash	*Aloe littoralis* Baker
äray	*Aloe* L.
argeesaa sodu	*Aloe calidophila* Reynolds
argeesaa	*Aloe* L.
	Aloe debrana Christian
	Aloe megalacantha Baker subsp. *megalacantha*

argeesaa (cont.) . . . . . . *Aloe rivae* Baker
Aloe trichosantha A.Berger subsp. *trichosantha*
Aloe yavellana Reynolds

argessa . . . . . . *Aloe megalacantha* Baker subsp. *megalacantha*

arret . . . . . . *Aloe percrassa* Tod.

asevar . . . . . . *Aloe succotrina* Weston

atzavara . . . . . . *Aloe arborescens* Mill. var. *arborescens*
Aloe maculata All.
Aloe vera (L.) Burm.f.

aukoreb . . . . . . *Aloe asperifolia* A.Berger
Aloe hereroensis Engl. var. *hereroensis*
Aloe littoralis Baker
Aloe zebrina Baker

azebra . . . . . . *Aloe arborescens* Mill. var. *arborescens*

azebre vegetal . . . . . . *Aloe vera* (L.) Burm.f.

azebre . . . . . . *Aloe vera* (L.) Burm.f.

ba toba xha . . . . . . *Aloe variegata* L.

baard-aalwyn . . . . . . *Aloe aristata* Haw.

babosa . . . . . . *Aloe arborescens* Mill. var. *arborescens*
Aloe vera (L.) Burm.f.

babosa-medicinal . . . . . . *Aloe vera* (L.) Burm.f.

balli nyibi . . . . . . *Aloe buettneri* A.Berger

balli nyiwa . . . . . . *Aloe buettneri* A.Berger

balsemera . . . . . . *Aloe arborescens* Mill. var. *arborescens*
Aloe maculata All.

bamalagba . . . . . . *Aloe buettneri* A.Berger

bangio fauru . . . . . . *Aloe buettneri* A.Berger

Barbados aloe . . . . . . *Aloe vera* (L.) Burm.f.

Barberton aalwyn . . . . . . *Aloe barbertoniae* Pole-Evans

Barberton aloe . . . . . . *Aloe barbertoniae* Pole-Evans

barolo . . . . . . *Aloe castanea* Schönland

Basil Christian's aloe . . . . . . *Aloe christianii* Reynolds

Basotoland aloe . . . . . . *Aloe polyphylla* Schönland ex Pillans

bastard quiver tree . . . . . . *Aloe pillansii* L.Guthrie

basteraalwyn . . . . . . *Aloe ×principis* (Haw.) Stearn

basterkokerboom . . . . . . *Aloe pillansii* L.Guthrie

baza . . . . . . *Aloe buettneri* A.Berger

beautiful aloe . . . . . . *Aloe speciosa* Baker

bec de perroquet . . . . . . *Aloe zebrina* Baker

belarmintz . . . . . . *Aloe arborescens* Mill. var. *arborescens*

berg alwyn . . . . . . *Aloe broomii* Schönland var. *broomii*

bergaalwee . . . . . *Aloe broomii* Schönland var. *broomii*
Aloe succotrina Weston

bergaalwyn . . . . . *Aloe broomii* Schönland var. *broomii*
Aloe ferox Mill.
Aloe gerstneri Reynolds
Aloe hereroensis Engl. var. *hereroensis*
Aloe khamiesensis Pillans
Aloe littoralis Baker
Aloe marlothii A.Berger subsp. *marlothii*
Aloe peglerae Schönland
Aloe plicatilis (L.) Mill.
Aloe pratensis Baker
Aloe reitzii Reynolds var. *reitzii*
Aloe succotrina Weston

beskore . . . . . *Aloe buettneri* A.Berger

bikakalubamba . . . . . *Aloe lateritia* Engl. var. *lateritia*

bindamutshe . . . . . *Aloe marlothii* A.Berger subsp. *marlothii*

bindamutsho . . . . . *Aloe marlothii* A.Berger subsp. *marlothii*

binda-mutsho . . . . . *Aloe marlothii* A.Berger subsp. *marlothii*

bito-xha . . . . . *Aloe vera* (L.) Burm.f.

bitter aloe . . . . . *Aloe ferox* Mill.
Aloe framesii L.Bolus

bitteraalwyn . . . . . *Aloe ferox* Mill.
Aloe framesii L.Bolus
Aloe marlothii A.Berger subsp. *marlothii*

bittervygie . . . . . *Aloe melanacantha* A.Berger

black thorn aloe . . . . . *Aloe melanacantha* A.Berger

blindamutsho . . . . . *Aloe marlothii* A.Berger subsp. *marlothii*

blou aalwyn . . . . . *Aloe glauca* Mill.

blouaalwee . . . . . *Aloe glauca* Mill.

blou-aalwee . . . . . *Aloe striata* Haw.

blouaalwyn . . . . . *Aloe glauca* Mill.
Aloe striata Haw.

blou-aalwyn . . . . . *Aloe striata* Haw.

blou-bontaalwyn . . . . . *Aloe verdoorniae* Reynolds

blou-kransaalwyn . . . . . *Aloe mutabilis* Pillans

blue krantz aloe . . . . . *Aloe mutabilis* Pillans

blue spotted aloe . . . . . *Aloe verdoorniae* Reynolds

boekaalwyn . . . . . *Aloe suprafoliata* Pole-Evans

boi . . . . . *Aloe buettneri* A.Berger

bole-siyah . . . . . *Aloe succotrina* Weston

bol-seoh . . . . . *Aloe littoralis* Baker

Bombay aloe . . . . . *Aloe succotrina* Weston

Bonda Ruwari aloe . . . . *Aloe cameronii* Hemsl. var. *bondana* Reynolds
bontaalwee . . . . *Aloe maculata* All.
Aloe zebrina Baker
bontaalwyn . . . . *Aloe affinis* A.Berger
Aloe davyana Schönland
Aloe grandidentata Salm-Dyck
Aloe krapohliana Marloth
Aloe maculata All.
Aloe prinslooi I.Verd. & D.S.Hardy
Aloe variegata L.
Aloe zebrina Baker
bontbees . . . . *Aloe variegata* L.
bontblaar aalwyn . . . . *Aloe davyana* Schönland
bontblaaraalwyn . . . . *Aloe davyana* Schönland
Aloe greatheadii Schönland
bont-o-t'korrie. . . . . *Aloe arenicola* Reynolds
book aloe. . . . . *Aloe suprafoliata* Pole-Evans
boom aalwyn . . . . *Aloe pillansii* L.Guthrie
boomaalwyn. . . . . *Aloe barberae* Dyer
Aloe marlothii A.Berger subsp. *marlothii*
boomalwyn . . . . *Aloe barberae* Dyer
borselaalwyn . . . . *Aloe rupestris* Baker
boskokerboom . . . . *Aloe dichotoma* Masson
Aloe ramosissima Pillans
botaalwee . . . . *Aloe zebrina* Baker
bottle brush aloe. . . . . *Aloe spicata* L.f.
bottlebrush aloe . . . . *Aloe rupestris* Baker
bottle-brush aloe. . . . . *Aloe rupestris* Baker
Branddraai aloe . . . . *Aloe branddraaiensis* Groenew.
Branddraai-aalwyn . . . . *Aloe branddraaiensis* Groenew.
broadleaf aloe. . . . . *Aloe maculata* All.
broad-leaved aloe . . . . *Aloe maculata* All.
broad-leaved grass aloe . . . . *Aloe boylei* Baker
broad-leaved yellow grass aloe . . . . *Aloe kraussii* Baker
bruinaalwyn . . . . *Aloe vryheidensis* Groenew.
Bullock's bottle brush aloe . . . . *Aloe taurii* L.C.Leach
Burgersfort aloe . . . . *Aloe burgersfortensis* Reynolds
Burgersfortaalwyn . . . . *Aloe burgersfortensis* Reynolds
Burgersfort-bontaalwyn . . . . *Aloe burgersfortensis* Reynolds
burn plant . . . . *Aloe vera* (L.) Burm.f.

cacto-dos-aflitos . . . . . . *Aloe vera* (L.) Burm.f.
Cameron's Ruwari aloe . . . . . . *Aloe cameronii* Hemsl. var. *cameronii*
Camper's aloe . . . . . . *Aloe camperi* Schweinf.
candelabra aloe . . . . . . *Aloe arborescens* Mill. var. *arborescens*
Aloe candelabrum A.Berger
candelabra plant . . . . . . *Aloe arborescens* Mill. var. *arborescens*
candelabra-plant . . . . . . *Aloe arborescens* Mill. var. *arborescens*
candelabrum aloe . . . . . . *Aloe candelabrum* A.Berger
cannon aloe . . . . . . *Aloe claviflora* Burch.
Cape aloe gel . . . . . . *Aloe ferox* Mill.
Cape aloe . . . . . . *Aloe ferox* Mill.
Cape aloes (bitter fraction) . . . . . . *Aloe ferox* Mill.
Cape prickly aloe . . . . . . *Aloe ferox* Mill.
caraguatá . . . . . . *Aloe vera* (L.) Burm.f.
carriapolum . . . . . . *Aloe littoralis* Baker
cat's-tail aloe . . . . . . *Aloe castanea* Schönland
cây aloe vera . . . . . . *Aloe vera* (L.) Burm.f.
cây lô hội . . . . . . *Aloe vera* (L.) Burm.f.
cây nha đam . . . . . . *Aloe vera* (L.) Burm.f.
Chabaud's aloe . . . . . . *Aloe chabaudii* Schönland var. *chabaudii*
chaudala . . . . . . *Aloe guerrae* Reynolds
chennanayakam . . . . . . *Aloe littoralis* Baker
Aloe succotrina Weston
chestnut brown aloe . . . . . . *Aloe castanea* Schönland
chhotakanvar . . . . . . *Aloe littoralis* Baker
chidzima mliro . . . . . . *Aloe* L.
chigiakia . . . . . . *Aloe* L.
Aloe chabaudii Schönland var. *chabaudii*
Aloe excelsa A.Berger var. *excelsa*
Aloe greatheadii Schönland
chikowa . . . . . . *Aloe* L.
Aloe chabaudii Schönland var. *chabaudii*
Aloe cryptopoda Baker
Aloe excelsa A.Berger var. *excelsa*
Aloe greatheadii Schönland
chinikala bunda . . . . . . *Aloe littoralis* Baker
chinikalabanda . . . . . . *Aloe littoralis* Baker
chinna kalabanda . . . . . . *Aloe vera* (L.) Burm.f.
chintembwe . . . . . . *Aloe* L.
Aloe arborescens Mill. var. *arborescens*

chinthembwe . . . . . . *Aloe* L.
Aloe bulbicaulis Christian
Aloe cameronii Hemsl. var. *cameronii*
Aloe cameronii Hemsl. var. *dedzana* Reynolds
Aloe canis S.Lane
Aloe chabaudii Schönland var. *chabaudii*
Aloe chabaudii Schönland var. *mlanjeana* Christian
Aloe christianii Reynolds
Aloe duckeri Christian
Aloe mawii Christian
Aloe zebrina Baker

chinungu . . . . . . *Aloe* L.
Aloe chabaudii Schönland var. *chabaudii*
Aloe excelsa A.Berger var. *excelsa*
Aloe greatheadii Schönland

chinyangami. . . . . . *Aloe* L.
Aloe chabaudii Schönland var. *chabaudii*
Aloe cryptopoda Baker
Aloe excelsa A.Berger var. *excelsa*
Aloe greatheadii Schönland

chirukattali. . . . . . *Aloe littoralis* Baker

chirukuttali. . . . . . *Aloe vera* (L.) Burm.f.

chisongwe . . . . . . *Aloe chabaudii* Schönland var. *chabaudii*

chitembwe. . . . . . *Aloe* L.
Aloe cryptopoda Baker
Aloe swynnertonii Rendle

chitseyse . . . . . . *Aloe arborescens* Mill. subsp. *arborescens*

chitupa . . . . . . *Aloe cryptopoda* Baker

chiwiriwiri . . . . . . *Aloe* L.
Aloe arborescens Mill. var. *arborescens*

Chizarira escarpment aloe . . . . . . *Aloe chabaudii* Schönland var. *chabaudii*

chizimamuliro. . . . . . *Aloe christianii* Reynolds

chizime . . . . . . *Aloe christianii* Reynolds

choje. . . . . . *Aloe dichotoma* Masson
Aloe variegata L.

chokokwet . . . . . . *Aloe secundiflora* Engl. var. *tweediae* (Christian) Wabuyele

chota-kunwar. . . . . . *Aloe littoralis* Baker

cilombo . . . . . . *Aloe chabaudii* Schönland var. *chabaudii*

cisongwe . . . . . . *Aloe chabaudii* Schönland var. *chabaudii*

citembwe. . . . . . *Aloe cryptopoda* Baker

citupa . . . . . . *Aloe cryptopoda* Baker

Clanwilliam aloe . . . . . . *Aloe comosa* Marloth & A.Berger

Clanwilliamaalwyn . . . . . . *Aloe comosa* Marloth & A.Berger

climbing aloe . . . . . . *Aloe ciliaris* Haw. var. *ciliaris*

climbing-flower aloe . . . . *Aloe suffulta* Reynolds
cloud-borne aloe . . . . *Aloe nubigena* Groenew.
coast aloe . . . . *Aloe thraskii* Baker
coastal aloe . . . . *Aloe vera* (L.) Burm.f.
Coega aloe . . . . *Aloe bowiea* Schult. & Schult.f.
Coega-aalwyn . . . . *Aloe bowiea* Schult. & Schult.f.
coiled aloe . . . . *Aloe polyphylla* Schönland ex Pillans
common aloe . . . . *Aloe ferox* Mill.
Aloe vera (L.) Burm.f.
common soap aloe . . . . *Aloe maculata* All.
Compton's aloe . . . . *Aloe comptonii* Reynolds
Cooper's aloe . . . . *Aloe cooperi* Baker subsp. *cooperi*
coral aloe . . . . *Aloe striata* Haw.
cultivated aloe . . . . *Aloe ferox* Mill.
cura-cancros . . . . *Aloe vera* (L.) Burm.f.
Curaçao aloe . . . . *Aloe vera* (L.) Burm.f.
curalotodo . . . . *Aloe maculata* All.

daar biyu . . . . *Aloe somaliensis* W.Watson var. *somaliensis*
daar burruk . . . . *Aloe retrospiciens* Reynolds
daar burruq . . . . *Aloe retrospiciens* Reynolds
daar burug . . . . *Aloe retrospiciens* Reynolds
daar der . . . . *Aloe eminens* Reynolds & P.R.O.Bally
Aloe gracilicaulis Reynolds & P.R.O.Bally
Aloe medishiana Reynolds & P.R.O.Bally
daar lebi . . . . *Aloe trichosantha* A.Berger subsp. *trichosantha*
daar merodi . . . . *Aloe rigens* Reynolds & P.R.O.Bally var. *rigens*
Aloe trichosantha A.Berger subsp. *trichosantha*
daar . . . . *Aloe* L.
Aloe parvidens M.G.Gilbert & Sebsebe
Aloe peckii P.R.O.Bally & I.Verd.
Aloe percrassa Tod.
Aloe pirottae A.Berger
Aloe scobinifolia Reynolds & P.R.O.Bally
Aloe trichosantha A.Berger subsp. *trichosantha*
dacar biyu . . . . *Aloe somaliensis* W.Watson var. *somaliensis*
dacar dheer . . . . *Aloe eminens* Reynolds & P.R.O.Bally
Aloe gracilicaulis Reynolds & P.R.O.Bally
dacar maroodi . . . . *Aloe rigens* Reynolds & P.R.O.Bally var. *rigens*
dacar qaraar . . . . *Aloe microdonta* Chiov.
dacar . . . . *Aloe parvidens* M.G.Gilbert & Sebsebe
Aloe ruspoliana Baker

De Wet's aloe . . . . *Aloe dewetii* Reynolds
degoree . . . . *Aloe variegata* L.
deurmekaarkoppie. . . . . . . *Aloe hereroensis* Engl. var. *hereroensis*
die dikke . . . . *Aloe dichotoma* Masson
die lange . . . . *Aloe pillansii* L.Guthrie
dilang-boaia . . . . *Aloe vera* (L.) Burm.f.
dilang-halo. . . . . . . *Aloe vera* (L.) Burm.f.
dilenga . . . . *Aloe nuttii* Baker
dishashanogha . . . . *Aloe esculenta* L.C.Leach
Aloe zebrina Baker
djimbelia . . . . *Aloe zebrina* Baker
dolomite aloe . . . . *Aloe vryheidensis* Groenew.
doringaalwyn . . . . *Aloe candelabrum* A.Berger
doringveldaalwyn . . . . *Aloe candelabrum* A.Berger
Dr Kirk's aloe . . . . *Aloe cryptopoda* Baker
duine-aalwyn . . . . *Aloe brevifolia* Mill. var. *brevifolia*
dune aloe. . . . . . . *Aloe thraskii* Baker
dwala aloe . . . . *Aloe chabaudii* Schönland var. *chabaudii*
dwarf hedgehog aloe . . . . *Aloe humilis* (L.) Mill.
dwarf yellow grass aloe . . . . *Aloe linearifolia* A.Berger
dyke aloe . . . . *Aloe ortholopha* Christian & Milne-Redh.

echichuviwa . . . . *Aloe kedongensis* Reynolds
Aloe lateritia Engl. var. *lateritia*
Aloe secundiflora Engl. var. *secundiflora*
echte aloe . . . . *Aloe vera* (L.) Burm.f.
echuchuka . . . . *Aloe kedongensis* Reynolds
Aloe secundiflora Engl. var. *secundiflora*
echuchuku . . . . *Aloe lateritia* Engl. var. *lateritia*
echuchukua . . . . *Aloe secundiflora* Engl. var. *secundiflora*
Aloe secundiflora Engl. var. *tweediae* (Christian) Wabuyele
Aloe wilsonii Reynolds
Ecklon's aloe . . . . *Aloe ecklonis* Salm-Dyck
Ecklon-se-aalwyn . . . . *Aloe ecklonis* Salm-Dyck
edundu . . . . *Aloe zebrina* Baker
eel aloe . . . . *Aloe alooides* (Bolus) Druten
ekundu . . . . *Aloe esculenta* L.C.Leach
Aloe zebrina Baker
eliya . . . . *Aloe succotrina* Weston
elva. . . . . . . *Aloe littoralis* Baker
Aloe succotrina Weston

elwa . . . . . *Aloe littoralis* Baker
Aloe vera (L.) Burm.f.
empofu . . . . . *Aloe tenuior* Haw.
endadaijoko . . . . . *Aloe secundiflora* Engl. var. *secundiflora*
endadaiyoku. . . . . *Aloe secundiflora* Engl. var. *secundiflora*
endobo . . . . . *Aloe esculenta* L.C.Leach
endombwe. . . . . *Aloe esculenta* L.C.Leach
erreh . . . . . *Aloe camperi* Schweinf.
Aloe percrassa Tod.
Aloe trichosantha A.Berger subsp. *trichosantha*
erva-azebra . . . . . *Aloe vera* (L.) Burm.f.
erva-babosa . . . . . *Aloe arborescens* Mill. var. *arborescens*
Aloe vera (L.) Burm.f.
erva-que-arde. . . . . *Aloe vera* (L.) Burm.f.
esugoroi. . . . . *Aloe secundiflora* Engl. var. *secundiflora*
esuguroi. . . . . *Aloe kedongensis* Reynolds
Aloe lateritia Engl. var. *lateritia*
etchuka . . . . . *Aloe secundiflora* Engl. var. *tweediae* (Christian) Wabuyele

fan aloe . . . . . *Aloe plicatilis* (L.) Mill.
fence aloe . . . . . *Aloe tenuior* Haw.
fiery climber. . . . . *Aloe ciliaris* Haw. var. *tidmarshii* Schönland
flat-flowered aloe . . . . . *Aloe marlothii* A.Berger subsp. *marlothii*
flor do deserto . . . . . *Aloe vera* (L.) Burm.f.
foguetes de Natal . . . . . *Aloe arborescens* Mill. var. *arborescens*
foguetes-de-Natal . . . . . *Aloe arborescens* Mill. var. *arborescens*
folonji . . . . . *Aloe rabaiensis* Rendle
Foster's aloe. . . . . *Aloe fosteri* Pillans
Fransaalwee. . . . . *Aloe pluridens* Haw.
Fransaalwyn. . . . . *Aloe pluridens* Haw.
Franschhoek aloe . . . . . *Aloe plicatilis* (L.) Mill.
Franschhoekaalwee. . . . . *Aloe plicatilis* (L.) Mill.
Franschhoekaalwyn. . . . . *Aloe plicatilis* (L.) Mill.
Franschoekaalwyn. . . . . *Aloe plicatilis* (L.) Mill.
French aloe . . . . . *Aloe pluridens* Haw.
French hoek aloe. . . . . *Aloe plicatilis* (L.) Mill.
fringing broader-leaved aloe . . . . . *Aloe ciliaris* Haw. var. *ciliaris*
fynbos grass aloe . . . . . *Aloe micracantha* Haw.
fynbosgrasaalwyn . . . . . *Aloe micracantha* Haw.

gabar . . . . . . *Aloe* L.
gaharn . . . . . . *Aloe littoralis* Baker
garaa . . . . . . *Aloe pluridens* Haw.
garab . . . . . . *Aloe dichotoma* Masson
garas . . . . . . *Aloe dichotoma* Masson
Gariep aloe . . . . . . *Aloe gariepensis* Pillans
gavakava . . . . . . *Aloe* L.
Aloe chabaudii Schönland var. *chabaudii*
Aloe excelsa A.Berger var. *excelsa*
Aloe globuligemma Pole-Evans
Aloe greatheadii Schönland
Aloe ortholopha Christian & Milne-Redh.
Aloe vera (L.) Burm.f.
gave wamtchanga . . . . . . *Aloe* L.
Aloe cryptopoda Baker
gavi . . . . . . *Aloe menyharthii* Baker subsp. *menyharthii*
Gazaland aloe . . . . . . *Aloe spicata* L.f.
gbadu . . . . . . *Aloe buettneri* A.Berger
geelaalwee . . . . . . *Aloe cryptopoda* Baker
Aloe pienaarii Pole-Evans
geelaalwyn . . . . . . *Aloe cryptopoda* Baker
Aloe pienaarii Pole-Evans
geelkransaalwyn . . . . . . *Aloe mutabilis* Pillans
genenoo . . . . . . *Aloe calidophila* Reynolds
Aloe camperi Schweinf.
Aloe trichosantha A.Berger subsp. *trichosantha*
Gerstner's aloe . . . . . . *Aloe gerstneri* Reynolds
ghikanvar . . . . . . *Aloe vera* (L.) Burm.f.
ghiu kumari . . . . . . *Aloe vera* (L.) Burm.f.
ghrita kumari . . . . . . *Aloe vera* (L.) Burm.f.
ghrit-kumari . . . . . . *Aloe vera* (L.) Burm.f.
giant quiver tree . . . . . . *Aloe pillansii* L.Guthrie
giant quiver-tree . . . . . . *Aloe pillansii* L.Guthrie
gilodu . . . . . . *Aloe zebrina* Baker
gladdeblaaraalwyn . . . . . . *Aloe striata* Haw.
godole uta . . . . . . *Aloe* L.
Aloe calidophila Reynolds
Aloe trichosantha A.Berger subsp. *trichosantha*
godzongo . . . . . . *Aloe* L.
Aloe chabaudii Schönland var. *chabaudii*
Aloe excelsa A.Berger var. *excelsa*
Aloe greatheadii Schönland
golonje . . . . . . *Aloe secundiflora* Engl. var. *secundiflora*

goree . . . . . *Aloe ferox* Mill.
Aloe melanacantha A.Berger
goreebosch . . . . . *Aloe ferox* Mill.
goresib . . . . . *Aloe littoralis* Baker
gorge de perdrix . . . . . *Aloe zebrina* Baker
grasaalwyn . . . . . *Aloe chortolirioides* A.Berger var. *chortolirioides*
Aloe chortolirioides A.Berger var. *woolliana* (Pole-Evans) Glen & D.S.Hardy
Aloe ecklonis Salm-Dyck
Aloe kniphofioifdes Baker
Aloe verecunda Pole-Evans
Graskop aloe . . . . . *Aloe alooides* (Bolus) Druten
Graskopaalwyn . . . . . *Aloe alooides* (Bolus) Druten
grass aloe . . . . . *Aloe cooperi* Baker subsp. *pulchra* Glen & D.S.Hardy
Aloe ecklonis Salm-Dyck
Aloe kniphofioides Baker
Aloe myriacantha (Haw.) Schult. & Schult.f.
Aloe verecunda Pole-Evans
Greathead's aloe . . . . . *Aloe greatheadii* Schönland
Greathead's spotted leaf aloe . . . . . *Aloe greatheadii* Schönland
Greatheads spotted-leaf aloe . . . . . *Aloe greatheadii* Schönland
green-sheathed narrow-leaved aloe . . . . . *Aloe tenuior* Haw.
grey aloe . . . . . *Aloe chabaudii* Schönland var. *chabaudii*
groenaalwyn . . . . . *Aloe camperi* Schweinf.
groenblomaalwyn . . . . . *Aloe viridiflora* Reynolds
grootaalwyn . . . . . *Aloe peglerae* Schönland
grootgrasaalwyn . . . . . *Aloe ecklonis* Salm-Dyck
grootrivieraalwyn . . . . . *Aloe gariepensis* Pillans
grysaalwyn . . . . . *Aloe chabaudii* Schönland var. *chabaudii*
guar patha . . . . . *Aloe vera* (L.) Burm.f.
guinea-fowl aloe . . . . . *Aloe aristata* Haw.
gweravana . . . . . *Aloe* L.
Aloe chabaudii Schönland var. *chabaudii*
Aloe excelsa A.Berger var. *excelsa*
Aloe greatheadii Schönland

hántsàr gííwáá . . . . . *Aloe buettneri* A.Berger
hantsar giwa . . . . . *Aloe buettneri* A.Berger
hargeisa sodu . . . . . *Aloe calidophila* Reynolds
hargeisa . . . . . *Aloe rabaiensis* Rendle
harguessa . . . . . *Aloe kedongensis* Reynolds
Aloe lateritia Engl. var. *lateritia*
Aloe secundiflora Engl. var. *secundiflora*

Hazel's rock aloe . . . . . *Aloe hazeliana* Reynolds var. *hazeliana*
hedgehog aloe . . . . . *Aloe humilis* (L.) Mill.
heejersaa . . . . . *Aloe calidophila* Reynolds
Aloe debrana Christian
Aloe megalacantha Baker subsp. *megalacantha*
Aloe yavellana Reynolds
heiningaalwyn . . . . . *Aloe striatula* Haw. var. *striatula*
heksekringe . . . . . *Aloe asperifolia* A.Berger
Herero aloe . . . . . *Aloe hereroensis* Engl. var. *hereroensis*
Hereroland aloe . . . . . *Aloe hereroensis* Engl. var. *hereroensis*
Hererolandaalwyn . . . . . *Aloe hereroensis* Engl. var. *hereroensis*
heuningaalwyn . . . . . *Aloe tenuior* Haw.
hierba del acíbar . . . . . *Aloe succotrina* Weston
hlaba . . . . . *Aloe* L.
Aloe ferox Mill.
hloho tsa makaka . . . . . *Aloe ecklonis* Salm-Dyck
Aloe kraussii Baker
hloho-tsa-makaka . . . . . *Aloe ecklonis* Salm-Dyck
Aloe kraussii Baker
Howman's cliff aloe . . . . . *Aloe hazeliana* Reynolds var. *howmanii* (Reynolds) S.Carter
hsiang tan . . . . . *Aloe vera* (L.) Burm.f.
humpets'k'in-ki . . . . . *Aloe vera* (L.) Burm.f.
Hunyani range aloe . . . . . *Aloe chabaudii* Schönland var. *chabaudii*

iandala . . . . . *Aloe zebrina* Baker
ibugubugu . . . . . *Aloe lateritia* Engl. var. *lateritia*
icena elikhulu . . . . . *Aloe pruinosa* Reynolds
icena lamatshe . . . . . *Aloe vanbalenii* Pillans
icena . . . . . *Aloe* L.
Aloe chabaudii Schönland var. *chabaudii*
Aloe dyeri Schönland
Aloe excelsa A.Berger var. *excelsa*
Aloe globuligemma Pole-Evans
Aloe greatheadii Schönland
Aloe greenii Baker
Aloe maculata All.
Aloe melanacantha A.Berger
Aloe mudenensis Reynolds
Aloe ortholopha Christian & Milne-Redh.
Aloe parvibracteata Schönland
Aloe pretoriensis Pole-Evans
Aloe pruinosa Reynolds
Aloe striata Haw.

icena (cont.) . . . *Aloe suprafoliata* Pole-Evans
Aloe umfoloziensis Reynolds

icenalamatshe . . . *Aloe vanbalenii* Pillans

icenandhlovu . . . *Aloe vanbalenii* Pillans

icenlandhlovu . . . *Aloe vanbalenii* Pillans

`iegedel fuga . . . *Aloe* L.

ifouaman . . . *Aloe longistyla* Baker
Aloe macrocarpa Tod. subsp. *macrocarpa*

igikakarubaamba . . . *Aloe lateritia* Engl. var. *lateritia*

igikakarubamba . . . *Aloe lateritia* Engl. var. *lateritia*

ihala . . . *Aloe maculata* All.

ihlaba . . . *Aloe chabaudii* Schönland var. *chabaudii*

ikala . . . *Aloe marlothii* A.Berger subsp. *marlothii*
Aloe spectabilis Reynolds

ikalana . . . *Aloe tenuior* Haw.

ikalene . . . *Aloe ciliaris* Haw. var. *ciliaris*

ikhala . . . *Aloe arborescens* Mill. subsp. *arborescens*
Aloe candelabrum A.Berger
Aloe ferox Mill.
Aloe marlothii A.Berger subsp. *marlothii*
Aloe speciosa Baker
Aloe thraskii Baker

ikhalana . . . *Aloe tenuior* Haw.

ikhalene . . . *Aloe tenuior* Haw.

ikshuramallika . . . *Aloe littoralis* Baker

ilicena lamatshe . . . *Aloe vanbalenii* Pillans

ilicena . . . *Aloe dyeri* Schönland
Aloe greenii Baker
Aloe maculata All.
Aloe melanacantha A.Berger
Aloe mudenensis Reynolds
Aloe parvibracteata Schönland
Aloe pretoriensis Pole-Evans
Aloe striata Haw.
Aloe suprafoliata Pole-Evans
Aloe umfoloziensis Reynolds

ilihala . . . *Aloe maculata* All.

ilikala . . . *Aloe marlothii* A.Berger subsp. *marlothii*
Aloe spectabilis Reynolds

ilva . . . *Aloe succotrina* Weston

imanga . . . *Aloe parvibracteata* Schönland

imangani . . . *Aloe excelsa* A.Berger var. *excelsa*

imboma . . . *Aloe* L.

imihlaba . . . . . *Aloe candelabrum* A.Berger
Aloe marlothii A.Berger subsp. *marlothii*
Aloe thraskii Baker
impondondo . . . . . *Aloe barberae* Dyer
incothobe . . . . . *Aloe boylei* Baker
Indian aloe . . . . . *Aloe vera* (L.) Burm.f.
indlabendlazi . . . . . *Aloe barberae* Dyer
ingarigari . . . . . *Aloe lateritia* Engl. var. *lateritia*
ingcelwane . . . . . *Aloe maculata* All.
Aloe striata Haw.
inhlaba empofu . . . . . *Aloe tenuior* Haw.
inhlaba yentaba . . . . . *Aloe decumbens* (Reynolds) van Jaarsv.
inhlaba . . . . . *Aloe affinis* A.Berger
Aloe arborescens Mill. var. *arborescens*
Aloe barberae Dyer
Aloe candelabrum A.Berger
Aloe chabaudii Schönland var. *chabaudii*
Aloe dewetii Reynolds
Aloe excelsa A.Berger var. *excelsa*
Aloe ferox Mill.
Aloe greatheadii Schönland
Aloe maculata All.
Aloe marlothii A.Berger subsp. *marlothii*
Aloe parvibracteata Schönland
Aloe rupestris Baker
Aloe spectabilis Reynolds
Aloe spicata L.f.
Aloe suprafoliata Pole-Evans
Aloe tenuior Haw.
Aloe thraskii Baker
inhlaba-encane . . . . . *Aloe arborescens* Mill. subsp. *arborescens*
inhlabane . . . . . *Aloe marlothii* A.Berger subsp. *marlothii*
inhlabanzhlazi . . . . . *Aloe rupestris* Baker
inhlahlwane . . . . . *Aloe vanbalenii* Pillans
inhlatjana . . . . . *Aloe kniphofioides* Baker
Aloe minima Baker
inhlazi . . . . . *Aloe arborescens* Mill. subsp. *arborescens*
inkalame . . . . . *Aloe arborescens* Mill. subsp. *arborescens*
inkalane encane . . . . . *Aloe arborescens* Mill. subsp. *arborescens*
inkalane unkulu . . . . . *Aloe barberae* Dyer
inkalane . . . . . *Aloe arborescens* Mill. subsp. *arborescens*
Aloe candelabrum A.Berger
Aloe chabaudii Schönland var. *chabaudii*
Aloe cooperi Baker subsp. *cooperi*
Aloe parvibracteata Schönland

inkalane (cont.) . . . . *Aloe rupestris* Baker
Aloe suprafoliata Pole-Evans
Aloe tenuior Haw.

inkalane-encane . . . . *Aloe arborescens* Mill. subsp. *arborescens*

inkalane-enkulu . . . . *Aloe barberae* Dyer

inkalene encane . . . . *Aloe arborescens* Mill. subsp. *arborescens*

inkhalane . . . . *Aloe rupestris* Baker

inkhuphuyana . . . . *Aloe linearifolia* A.Berger

inkuphuyana . . . . *Aloe linearifolia* A.Berger

inocelwane . . . . *Aloe maculata* All.

intelezi . . . . *Aloe ciliaris* Haw. var. *ciliaris*
Aloe tenuior Haw.

Inyanga aloe . . . . *Aloe inyangensis* Christian var. *inyangensis*

iposo . . . . *Aloe arborescens* Mill. subsp. *arborescens*

iratune . . . . *Aloe volkensii* Engl. subsp. *volkensii*

iret matiso . . . . *Aloe percrassa* Tod.

iret . . . . *Aloe adigratana* Reynolds
Aloe camperi Schweinf.
Aloe percrassa Tod.

isihlabana . . . . *Aloe gerstneri* Reynolds

isihlabane . . . . *Aloe gerstneri* Reynolds

isiphukhutshane . . . . *Aloe minima* Baker

isiphukuthwane . . . . *Aloe boylei* Baker
Aloe cooperi Baker subsp. *cooperi*
Aloe ecklonis Salm-Dyck
Aloe minima Baker

isiphukutwane . . . . *Aloe boylei* Baker
Aloe cooperi Baker subsp. *cooperi*

isiphuthumana . . . . *Aloe cooperi* Baker subsp. *cooperi*

isiphuthumane . . . . *Aloe ecklonis* Salm-Dyck
Aloe boylei Baker
Aloe cooperi Baker subsp. *cooperi*
Aloe kraussii Baker

isipukushane . . . . *Aloe minima* Baker

isipukutwane . . . . *Aloe cooperi* Baker subsp. *cooperi*
Aloe ecklonis Salm-Dyck
Aloe kraussii Baker
Aloe minima Baker

isiputuma . . . . *Aloe minima* Baker

isiputumane . . . . *Aloe boylei* Baker
Aloe cooperi Baker subsp. *cooperi*
Aloe ecklonis Salm-Dyck

isisphukhutwane . . . . *Aloe kraussii* Baker

isisphukuthwane . . . . *Aloe ecklonis* Salm-Dyck
itembushia . . . . *Aloe greatheadii* Schönland
iwani . . . . *Aloe* L.
Aloe christianii Reynolds
iwata . . . . *Aloe nuttii* Baker

Jaffarabad aloe . . . . *Aloe vera* (L.) Burm.f.
jakkalsstert . . . . *Aloe claviflora* Burch.
jaya jaya . . . . *Aloe vera* (L.) Burm.f.
je'awiyon . . . . *Aloe jawiyon* Christie, Hannon & Oakman
jejeje . . . . *Aloe cooperi* Baker subsp. *pulchra* Glen & D.S.Hardy
John Ball's cliff aloe . . . . *Aloe ballii* Reynolds var. *ballii*
joloji . . . . *Aloe secundiflora* Engl. var. *secundiflora*
jolonji . . . . *Aloe kedongensis* Reynolds
Aloe lateritia Engl. var. *lateritia*
Aloe rabaiensis Rendle

Kaapse aalwyn . . . . *Aloe ferox* Mill.
kábàr gíïwáá . . . . *Aloe buettneri* A.Berger
kabar giwa . . . . *Aloe buettneri* A.Berger
kabargiwa . . . . *Aloe buettneri* A.Berger
kadio . . . . *Aloe buettneri* A.Berger
kakamamba . . . . *Aloe abyssinica* Lam. (exact application unknown)
kakaruamba . . . . *Aloe* L.
kakarutanga . . . . *Aloe dawei* A.Berger
kakruamba . . . . *Aloe* L.
kalabanda . . . . *Aloe vera* (L.) Burm.f.
kalaboel . . . . *Aloe littoralis* Baker
kalandoy . . . . *Aloe sinkatana* Reynolds
kalu-bolam . . . . *Aloe succotrina* Weston
Kamiesberg aalwyn . . . . *Aloe khamiesensis* Pillans
Kamiesberg aloe . . . . *Aloe khamiesensis* Pillans
kamingaminga . . . . *Aloe christianii* Reynolds
kandelaaraalwyn . . . . *Aloe candelabrum* A.Berger
kandio . . . . *Aloe buettneri* A.Berger
kanembe . . . . *Aloe* L.
Aloe zebrina Baker
kanniedood aloe . . . . *Aloe variegata* L.
Aloe zebrina Baker

kanniedood . . . . *Aloe ferox* Mill.
Aloe grandidentata Salm-Dyck
Aloe variegata L.
kanonaalwyn . . . . *Aloe claviflora* Burch.
kanya . . . . *Aloe littoralis* Baker
Kapaloë . . . . *Aloe ferox* Mill.
karangheri . . . . *Aloe babatiensis* Christian & I.Verd.
karia polam . . . . *Aloe littoralis* Baker
kariambolam . . . . *Aloe littoralis* Baker
karibolam . . . . *Aloe succotrina* Weston
kariya-polam . . . . *Aloe succotrina* Weston
Karoo aloe . . . . *Aloe longistyla* Baker
Karoo-aalwyn . . . . *Aloe ferox* Mill.
Aloe longistyla Baker
Aloe microstigma Salm-Dyck
katstert . . . . *Aloe claviflora* Burch.
katstertaalwyn . . . . *Aloe castanea* Schönland
Aloe claviflora Burch.
kattalai . . . . *Aloe vera* (L.) Burm.f.
kattavala . . . . *Aloe littoralis* Baker
Keimoesaalwyn . . . . *Aloe gariepensis* Pillans
kghopa . . . . *Aloe marlothii* A.Berger subsp. *marlothii*
kgopalmabalamantsi . . . . *Aloe zebrina* Baker
kgopane . . . . *Aloe davyana* Schönland
kgope . . . . *Aloe zebrina* Baker
kharatsa . . . . *Aloe polyphylla* Schönland ex Pillans
khauzi . . . . *Aloe menyharthii* Baker subsp. *menyharthii*
khondje . . . . *Aloe christianii* Reynolds
khonje wa fisi . . . . *Aloe chabaudii* Schönland var. *chabaudii*
khonje wa mbuzi . . . . *Aloe chabaudii* Schönland var. *chabaudii*
khonje wa mtchire . . . . *Aloe* L.
khopha . . . . *Aloe marlothii* A.Berger subsp. *marlothii*
khudsi . . . . *Aloe menyharthii* Baker subsp. *menyharthii*
khudzi . . . . *Aloe menyharthii* Baker subsp. *menyharthii*
khuzi . . . . *Aloe* L.
Aloe bulbicaulis Christian
Aloe cameronii Hemsl. var. *cameronii*
Aloe cameronii Hemsl. var. *dedzana* Reynolds
Aloe canis S.Lane
Aloe chabaudii Schönland var. *chabaudii*
Aloe chabaudii Schönland var. *mlanjeana* Christian
Aloe christianii Reynolds

khuzi (cont.) . . . . . *Aloe duckeri* Christian
Aloe mawii Christian
Aloe menyharthii Baker subsp. *menyharthii*

kibela . . . . . *Aloe buettneri* A.Berger

Kidachi aloe . . . . . *Aloe arborescens* Mill. subsp. *arborescens*

Kidachi Japanese aloe . . . . . *Aloe arborescens* Mill. subsp. *arborescens*

kidata . . . . . *Aloe lateritia* Engl. var. *lateritia*

kigagi . . . . . *Aloe lateritia* Engl. var. *lateritia*

kigaka . . . . . *Aloe kedongensis* Reynolds
Aloe lateritia Engl. var. *lateritia*
Aloe secundiflora Engl. var. *secundiflora*

kikaka . . . . . *Aloe lateritia* Engl. var. *lateritia*

kikakalubamba . . . . . *Aloe littoralis* Baker

kikakalumbamba . . . . . *Aloe lateritia* Engl. var. *lateritia*

kikalanga kibela . . . . . *Aloe buettneri* A.Berger

kikalangu . . . . . *Aloe buettneri* A.Berger

kikalangu-kibela . . . . . *Aloe paedogona* A.Berger

kikoli . . . . . *Aloe leptosiphon* A.Berger

kiluma . . . . . *Aloe kedongensis* Reynolds
Aloe lateritia Engl. var. *lateritia*
Aloe secundiflora Engl. var. *secundiflora*

kipapa . . . . . *Aloe ballyi* Reynolds

kirumi . . . . . *Aloe lateritia* Engl. var. *lateritia*
Aloe secundiflora Engl. var. *secundiflora*

kirumu . . . . . *Aloe kedongensis* Reynolds

kisikmamleo . . . . . *Aloe kedongensis* Reynolds

kisimamleao . . . . . *Aloe kilifiensis* Christian

kisimamleo . . . . . *Aloe lateritia* Engl. var. *lateritia*
Aloe nuttii Baker
Aloe secundiflora Engl. var. *secundiflora*

kisimando . . . . . *Aloe kedongensis* Reynolds
Aloe lateritia Engl. var. *lateritia*
Aloe secundiflora Engl. var. *secundiflora*

kitembo . . . . . *Aloe lateritia* Engl. var. *lateritia*

kithapa . . . . . *Aloe lateritia* Engl. var. *lateritia*

kitori . . . . . *Aloe kedongensis* Reynolds
Aloe lateritia Engl. var. *lateritia*
Aloe secundiflora Engl. var. *secundiflora*

kizimabupia . . . . . *Aloe greatheadii* Schönland

kizima-bupia . . . . . *Aloe greatheadii* Schönland

kizimlo . . . . . *Aloe rabaiensis* Rendle

klein bergaalwyn . . . . . *Aloe melanacantha* A.Berger

kleinaalwyn . . . .*Aloe bowiea* Schult. & Schult.f.
Aloe brevifolia Mill. var. *brevifolia*
Aloe davyana Schönland
Aloe greatheadii Schönland
Aloe mudenensis Reynolds
Aloe pruinosa Reynolds

kleinbergaalwee . . . . *Aloe melanacantha* A.Berger

kleinbergaalwyn . . . . *Aloe melanacantha* A.Berger

kleinbontaalwyn . . . .*Aloe grandidentata* Salm-Dyck

kleingrasaalwyn . . . .*Aloe thompsoniae* Groenew.

Kleinkaroo-aalwyn. . . . . *Aloe comptonii* Reynolds

klompiesaalwyn . . . .*Aloe kraussii* Baker

knoppies aalwyn. . . . .*Aloe globuligemma* Pole-Evans

knoppiesaalwyn . . . .*Aloe aculeata* Pole-Evans
Aloe globuligemma Pole-Evans

köcherbaum . . . . *Aloe dichotoma* Masson

kokerbaum . . . . *Aloe dichotoma* Masson

kokerboom . . . . *Aloe dichotoma* Masson
Aloe plicatilis (L.) Mill.

kokorutanga . . . . *Aloe dawei* A.Berger

komaree. . . . . *Aloe littoralis* Baker

koraalaalwyn . . . . *Aloe striata* Haw.

koreb. . . . . *Aloe hereroensis* Engl. var. *hereroensis*

korphad . . . . *Aloe vera* (L.) Burm.f.

kpipiko . . . . *Aloe buettneri* A.Berger

kraalaalwee . . . . *Aloe rupestris* Baker

kraalaalwyn . . . . *Aloe asperifolia* A.Berger
Aloe claviflora Burch.
Aloe ferox Mill.
Aloe pachygaster Dinter
Aloe rupestris Baker

kraalaloe . . . . *Aloe claviflora* Burch.

krans aloe . . . . *Aloe arborescens* Mill. subsp. *arborescens*

kransaalwee. . . . . *Aloe perfoliata* L.

kransaalwyn. . . . . *Aloe arborescens* Mill. subsp. *arborescens*
Aloe mutabilis Pillans
Aloe perfoliata L.

krantz aloe . . . . *Aloe arborescens* Mill. subsp. *arborescens*

Krapohl's aloe. . . . . *Aloe krapohliana* Marloth

Krapohl-se-aalwyn . . . . *Aloe krapohliana* Marloth

krimpvarkie . . . . *Aloe erinacea* D.S.Hardy

krimpvarkieaalwyn . . . . *Aloe humilis* (L.) Mill.

kroonaalwee. . . . .*Aloe polyphylla* Schönland ex Pillans

kroonaalwyn . . . . . . *Aloe polyphylla* Schönland ex Pillans
kumaree . . . . . . *Aloe vera* (L.) Burm.f.
kumari . . . . . . *Aloe littoralis* Baker
Aloe vera (L.) Burm.f.
kunvar . . . . . . *Aloe vera* (L.) Burm.f.
kxophane . . . . . . *Aloe davyana* Schönland
Aloe greatheadii Schönland

lääkeaaloe . . . . . . *Aloe vera* (L.) Burm.f.
lace aloe . . . . . . *Aloe aristata* Haw.
lægealoe . . . . . . *Aloe vera* (L.) Burm.f.
laeraalwyn . . . . . . *Aloe claviflora* Burch.
lahani kumari . . . . . . *Aloe littoralis* Baker
lai . . . . . . *Aloe cryptopoda* Baker
langnaaldaalwyn . . . . . . *Aloe aristata* Haw.
lank'u . . . . . . *Aloe vera* (L.) Burm.f.
lap'i . . . . . . *Aloe vera* (L.) Burm.f.
Lebombo aloe . . . . . . *Aloe spicata* L.f.
Lebomboaalwyn . . . . . . *Aloe spicata* L.f.
Lebombo-aalwyn . . . . . . *Aloe spicata* L.f.
legno aloe . . . . . . *Aloe vera* (L.) Burm.f.
lekhala kharatsa . . . . . . *Aloe polyphylla* Schönland ex Pillans
lekhala la Quthing . . . . . . *Aloe ferox* Mill.
lekhala le thaba . . . . . . *Aloe maculata* All.
lekhala qhalane . . . . . . *Aloe pratensis* Baker
lekhala qhalene . . . . . . *Aloe pratensis* Baker
lekhala . . . . . . *Aloe* L.
Aloe ecklonis Salm-Dyck
Aloe ferox Mill.
Aloe maculata All.
Aloe polyphylla Schönland ex Pillans
Aloe striatula Haw. var. *caesia* Reynolds
lekhala-la-Lesotho . . . . . . *Aloe maculata* All.
lekhala-la-Linakeng . . . . . . *Aloe pratensis* Baker
lekhala-la-Quthing . . . . . . *Aloe ferox* Mill.
lekhala-la-thaba . . . . . . *Aloe maculata* All.
Aloe polyphylla Schönland ex Pillans
lekhala-le-leholo . . . . . . *Aloe ferox* Mill.
lekhala-le-lenyenyane . . . . . . *Aloe aristata* Haw.
lekhalana . . . . . . *Aloe aristata* Haw.
Aloe ecklonis Salm-Dyck
Aloe kraussii Baker

lekhalana-la-balimo	*Aloe aristata* Haw.
lekhala-qhalane	*Aloe pratensis* Baker
lekopane	*Aloe globuligemma* Pole-Evans
lekxala-lathaba	*Aloe maculata* All.
lexalana	*Aloe kraussii* Baker
lichingwe	*Aloe christianii* Reynolds
lichongwe	*Aloe* L.
	Aloe bulbicaulis Christian
	Aloe cameronii Hemsl. var. *cameronii*
	Aloe cameronii Hemsl. var. *dedzana* Reynolds
	Aloe canis S.Lane
	Aloe chabaudii Schönland var. *chabaudii*
	Aloe chabaudii Schönland var. *mlanjeana* Christian
	Aloe cryptopoda Baker
	Aloe duckeri Christian
	Aloe mawii Christian
	Aloe swynnertonii Rendle
	Aloe zebrina Baker
licongwe	*Aloe christianii* Reynolds
	Aloe cryptopoda Baker
lidah buaya	*Aloe ferox* Mill.
lihlala	*Aloe maculata* All.
	Aloe vanbalenii Pillans
likokho	*Aloe dawei* A.Berger
lilinakha	*Aloe dawei* A.Berger
Limpopo spotted aloe	*Aloe longibracteata* Pole-Evans
Limpopobontaalwyn	*Aloe longibracteata* Pole-Evans
linakha	*Aloe volkensii* Engl. subsp. *volkensii*
lináloe	*Aloe vera* (L.) Burm.f.
lined red-spined aloe	*Aloe lineata* (Aiton) Haw. var. *lineata*
lineke	*Aloe kedongensis* Reynolds
	Aloe lateritia Engl. var. *lateritia*
	Aloe secundiflora Engl. var. *secundiflora*
liperege	*Aloe lateritia* Engl. var. *lateritia*
lishashankogha	*Aloe esculenta* L.C.Leach
	Aloe zebrina Baker
lisheshelu	*Aloe cooperi* Baker subsp. *cooperi*
lissigiri	*Aloe lateritia* Engl. var. *lateritia*
litembwetembwe	*Aloe lateritia* Engl. var. *lateritia*
llai	*Aloe chabaudii* Schönland var. *chabaudii*
lolisara	*Aloe vera* (L.) Burm.f.
long-awned aloe	*Aloe aristata* Haw.
losa	*Aloe* L.

loto do deserto . . . . . *Aloe vera* (L.) Burm.f.
Lowveld spotted aloe . . . . . *Aloe parvibracteata* Schönland
Lowveld spotted leaf aloe . . . . . *Aloe parvibracteata* Schönland
lu hui . . . . . *Aloe vera* (L.) Burm.f.
Luanda tree aloe . . . . . *Aloe littoralis* Baker
Luckhoffaalwyn . . . . . *Aloe variegata* L.
Luckhoffse aalwyn . . . . . *Aloe variegata* L.
lugaka . . . . . *Aloe* L.
luza . . . . . *Aloe flexilifolia* Christian
lyusi . . . . . *Aloe lateritia* Engl. var. *lateritia*

madaka . . . . . *Aloe chabaudii* Schönland var. *chabaudii*
magno gu dondialé . . . . . *Aloe buettneri* A.Berger
maguey morado . . . . . *Aloe vera* (L.) Burm.f.
maiden quiver tree . . . . . *Aloe ramosissima* Pillans
maiden's quiver tree . . . . . *Aloe ramosissima* Pillans
maiden's quiver-tree . . . . . *Aloe ramosissima* Pillans
mak bontaalwyn . . . . . *Aloe maculata* All.
makaalwyn . . . . . *Aloe ferox* Mill.
Aloe marlothii A.Berger subsp. *marlothii*
Aloe striata Haw.
makbontaalwyn . . . . . *Aloe* ×*schimperi* Tod.
mala . . . . . *Aloe* L.
Malapati aloe . . . . . *Aloe lutescens* Groenew.
maluwa . . . . . *Aloe buchananii* Baker
manasvato . . . . . *Aloe* L.
manga . . . . . *Aloe parvibracteata* Schönland
mangana grande . . . . . *Aloe barberae* Dyer
Aloe marlothii A.Berger subsp. *marlothii*
mangana . . . . . *Aloe chabaudii* Schönland var. *chabaudii*
Aloe zebrina Baker
mangani . . . . . *Aloe chabaudii* Schönland var. *chabaudii*
mangesa . . . . . *Aloe menyharthii* Baker subsp. *menyharthii*
mantombo . . . . . *Aloe zebrina* Baker
manyesa . . . . . *Aloe* L.
Aloe menyharthii Baker subsp. *menyharthii*
many-toothed aloe . . . . . *Aloe pluridens* Haw.
many-toothed tree aloe . . . . . *Aloe pluridens* Haw.
maposo . . . . . *Aloe buettneri* A.Berger

maroba-lihale . . . . . . *Aloe ecklonis* Salm-Dyck
Aloe kraussii Baker
maroba-lilale . . . . . . *Aloe ecklonis* Salm-Dyck
maroba-lithatle . . . . . . *Aloe ecklonis* Salm-Dyck
matiso . . . . . . *Aloe* L.
mazambron marron . . . . . . *Aloe macra* Haw.
Aloe purpurea Lam.
mazambron sauvage . . . . . . *Aloe macra* Haw.
Aloe purpurea Lam.
mazambron . . . . . . *Aloe vera* (L.) Burm.f.
mbudyadya . . . . . . *Aloe greatheadii* Schönland
mdyang'oma . . . . . . *Aloe* L.
Aloe cryptopoda Baker
Aloe menyharthii Baker subsp. *menyharthii*
meadow aloe . . . . . . *Aloe pratensis* Baker
medicinal aloe . . . . . . *Aloe ferox* Mill.
Aloe vera (L.) Burm.f.
medicirula-azebre . . . . . . *Aloe vera* (L.) Burm.f.
medisyneaalwyn . . . . . . *Aloe vera* (L.) Burm.f.
Mediterranean aloe . . . . . . *Aloe vera* (L.) Burm.f.
men-tipa . . . . . . *Aloe buettneri* A.Berger
merarie . . . . . . *Aloe ankobernesis* M.G.Gilbert & Sebsebe
Aloe debrana Christian
mgaka . . . . . . *Aloe* L.
mhanga yikulu . . . . . . *Aloe marlothii* A.Berger subsp. *marlothii*
mhanga . . . . . . *Aloe marlothii* A.Berger subsp. *marlothii*
Aloe parvibracteata Schönland
mhangani . . . . . . *Aloe* L.
Aloe excelsa A.Berger var. *excelsa*
Aloe greatheadii Schönland
mikaalwyn . . . . . . *Aloe barberae* Dyer
miniature aloe . . . . . . *Aloe krapohliana* Marloth
mist belt tree aloe . . . . . . *Aloe pretoriensis* Pole-Evans
mitre aloe . . . . . . *Aloe perfoliata* L.
mkakruamba . . . . . . *Aloe* L.
mlalangao . . . . . . *Aloe lateritia* Engl. var. *lateritia*
Aloe macrocarpa Tod. subsp. *wollastonii* (Rendle) Wabuyele
mocha aloe . . . . . . *Aloe succotrina* Weston
moda . . . . . . *Aloe buettneri* A.Berger
mogopa . . . . . . *Aloe marlothii* A.Berger subsp. *marlothii*
mohalakane . . . . . . *Aloe ferox* Mill.
Aloe striatula Haw. var. *caesia* Reynolds
Aloe striatula Haw. var. *striatula*

mokgopa . . . . *Aloe marlothii* A.Berger subsp. *marlothii*
mokqala. . . . . *Aloe spectabilis* Reynolds
móódáá . . . . *Aloe buettneri* A.Berger
mopane aloe . . . . *Aloe littoralis* Baker
mopane-aalwyn . . . . *Aloe littoralis* Baker
moshabbar. . . . . *Aloe succotrina* Weston
Mossel Bay hybrid aloe . . . . *Aloe ×principis* (Haw.) Stearn
mountain aloe . . . . *Aloe broomii* Schönland var. *broomii*
Aloe marlothii A.Berger subsp. *marlothii*
mountain bush aloe. . . . . *Aloe arborescens* Mill. subsp. *arborescens*
mpagana . . . . *Aloe marlothii* A.Berger subsp. *marlothii*
mradune . . . . *Aloe volkensii* Engl. subsp. *volkensii*
mratune. . . . . *Aloe lateritia* Engl. var. *lateritia*
msht `iegedel fuga . . . . *Aloe pulcherrima* M.G.Gilbert & Sebsebe
mshubili . . . . *Aloe nuttii* Baker
msubili . . . . *Aloe nuttii* Baker
mubudyadya . . . . *Aloe greatheadii* Schönland
Muden aloe . . . . *Aloe mudenensis* Reynolds
mũgwanũgũ . . . . *Aloe secundiflora* Engl. var. *secundiflora*
mukumi. . . . . *Aloe kedongensis* Reynolds
Aloe lateritia Engl. var. *lateritia*
Aloe secundiflora Engl. var. *secundiflora*
Munch's great Chimanimani aloe . . . . *Aloe munchii* Christian
mundumba . . . . *Aloe excelsa* A.Berger var. *excelsa*
musabar . . . . *Aloe littoralis* Baker
musabbar. . . . . *Aloe succotrina* Weston
musambar . . . . *Aloe littoralis* Baker
musambaran . . . . *Aloe littoralis* Baker
musanbar. . . . . *Aloe succotrina* Weston
Musapa aloe. . . . . *Aloe musapana* Reynolds
mushabhir . . . . *Aloe littoralis* Baker
mushambaram . . . . *Aloe littoralis* Baker
Aloe succotrina Weston

n||cru. . . . . *Aloe littoralis* Baker
Aloe zebrina Baker
n||uru. . . . . *Aloe zebrina* Baker
nahani kanvar. . . . . *Aloe littoralis* Baker
Namakwa-aalwyn . . . . *Aloe khamiesensis* Pillans

namanyesa . . . . *Aloe* L.
Aloe menyharthii Baker subsp. *menyharthii*
Namaqua aloe . . . . *Aloe khamiesensis* Pillans
Namib aalwyn . . . . *Aloe namibensis* Giess
Namib aloe . . . . *Aloe namibensis* Giess
na-pug-maande . . . . *Aloe buettneri* A.Berger
nasi . . . . *Aloe menyharthii* Baker subsp. *menyharthii*
Aloe swynnertonii Rendle
Natal aloe . . . . *Aloe spectabilis* Reynolds
Natalaalwyn . . . . *Aloe spectabilis* Reynolds
Natalse doringveldaalwyn . . . . *Aloe candelabrum* A.Berger
nemba . . . . *Aloe* L.
Aloe chabaudii Schönland var. *chabaudii*
new aloes . . . . *Aloe ferox* Mill.
ngafane . . . . *Aloe cryptopoda* Baker
Aloe wickensii Pole-Evans var. *wickensii*
ngaka . . . . *Aloe* L.
Aloe lateritia Engl. var. *lateritia*
ngarare . . . . *Aloe lateritia* Engl. var. *lateritia*
ngirya . . . . *Aloe nuttii* Baker
ngopa nara . . . . *Aloe marlothii* A.Berger subsp. *marlothii*
ngopane . . . . *Aloe aculeata* Pole-Evans
ngopani . . . . *Aloe aculeata* Pole-Evans
ngosiya . . . . *Aloe chabaudii* Schönland var. *chabaudii*
nibeets . . . . *Aloe nuttii* Baker
nimbéléké . . . . *Aloe buettneri* A.Berger
njandola . . . . *Aloe zebrina* Baker
nkaka . . . . *Aloe bukobana* Reynolds
nkalane . . . . *Aloe cooperi* Baker subsp. *cooperi*
Aloe rupestris Baker
nl/'ho'oru . . . . *Aloe zebrina* Baker
nllhoq'uru . . . . *Aloe angolensis* Baker
Aloe zebrina Baker
nllhoq'ùrù . . . . *Aloe angolensis* Baker
Aloe zebrina Baker
no hui . . . . *Aloe vera* (L.) Burm.f.
nooienskokerboom . . . . *Aloe ramosissima* Pillans
nsenjere . . . . *Aloe chabaudii* Schönland var. *chabaudii*
Aloe christianii Reynolds
nsesareso abrobe . . . . *Aloe buettneri* A.Berger
ntehiseng . . . . *Aloe aristata* Haw.

nyakaryayata . . . . *Aloe myriacantha* (Haw.) Schult. & Schult.f.

ober . . . . *Aloe* L.
ocothobe . . . . *Aloe boylei* Baker
octopus plant . . . . *Aloe arborescens* Mill. subsp. *arborescens*
octopus-plant . . . . *Aloe arborescens* Mill. subsp. *arborescens*
ödağacı . . . . *Aloe vera* (L.) Burm.f.
ogara . . . . *Aloe kedongensis* Reynolds
Aloe lateritia Engl. var. *lateritia*
Aloe secundiflora Engl. var. *secundiflora*
ojinkalangua . . . . *Aloe zebrina* Baker
okandala-kasengue . . . . *Aloe zebrina* Baker
okandala-kazengue . . . . *Aloe zebrina* Baker
okandole . . . . *Aloe palmiformis* Baker
Aloe zebrina Baker
okandolle . . . . *Aloe catengiana* Reynolds
Oldenland's bush aloe . . . . *Aloe arborescens* Mill. subsp. *arborescens*
olkos . . . . *Aloe kedongensis* Reynolds
Aloe lateritia Engl. var. *lateritia*
Aloe secundiflora Engl. var. *secundiflora*
omakundu . . . . *Aloe zebrina* Baker
omandobo . . . . *Aloe esculenta* L.C.Leach
omvi . . . . *Aloe buettneri* A.Berge
omwi . . . . *Aloe buettneri* A.Berger
oolowaton . . . . *Aloe littoralis* Baker
oorbeweidingsaalwyn . . . . *Aloe davyana* Schönland
opregte aalwyn . . . . *Aloe ferox* Mill.
opregte-aalwyn . . . . *Aloe ferox* Mill.
Aloe marlothii A.Berger subsp. *marlothii*
Orange River aloe . . . . *Aloe gariepensis* Pillans
Oranjerivier-aalwyn . . . . *Aloe gariepensis* Pillans
os suguroi . . . . *Aloe volkensii* Engl. subsp. *volkensii*
osuguroi . . . . *Aloe kedongensis* Reynolds
Aloe secundiflora Engl. var. *secundiflora*
Aloe volkensii Engl. subsp. *volkensii*
osuguru . . . . *Aloe kedongensis* Reynolds
otchandala-ekundu . . . . *Aloe zebrina* Baker
otchyandala . . . . *Aloe zebrina* Baker
otjindombo . . . . *Aloe hereroensis* Engl. var. *hereroensis*
Aloe littoralis Baker
Aloe zebrina Baker

otyiandola . . . . . . *Aloe zebrina* Baker
oviandala . . . . . . *Aloe zebrina* Baker

painted-leaved aloe . . . . . . *Aloe krapohliana* Marloth
partridge aloe . . . . . . *Aloe variegata* L.
partridge breast aloe . . . . . . *Aloe variegata* L.
Aloe zebrina Baker
partridge-breast aloe . . . . . . *Aloe variegata* L.
Pearson's aloe . . . . . . *Aloe pearsonii* Schönland
Pearson-se-aalwyn . . . . . . *Aloe pearsonii* Schönland
penca sábila . . . . . . *Aloe vera* (L.) Burm.f.
peria karalai . . . . . . *Aloe littoralis* Baker
pers-bontaalwyn . . . . . . *Aloe parvibracteata* Schönland
pets'in-ki . . . . . . *Aloe vera* (L.) Burm.f.
phurumela . . . . . . *Aloe polyphylla* Schönland ex Pillans
pikwe . . . . . . *Aloe christianii* Reynolds
Pillans' aloe . . . . . . *Aloe pillansii* L.Guthrie
pinangru . . . . . . *Aloe buettneri* A.Berger
pita zabila . . . . . . *Aloe vera* (L.) Burm.f.
pitazabila . . . . . . *Aloe maculata* All.
Aloe vera (L.) Burm.f.
pitera amarelo . . . . . . *Aloe vera* (L.) Burm.f.
planta-dos-milagres . . . . . . *Aloe vera* (L.) Burm.f.
planta-mistério . . . . . . *Aloe vera* (L.) Burm.f.
planta-que-cura . . . . . . *Aloe vera* (L.) Burm.f.
Plowes' grass aloe . . . . . . *Aloe plowesii* Reynolds
poeieraalwyn . . . . . . *Aloe pruinosa* Reynolds
pohon gaharu . . . . . . *Aloe vera* (L.) Burm.f.
pomo . . . . . . *Aloe nuttii* Baker
powder aloe . . . . . . *Aloe pruinosa* Reynolds
pozo . . . . . . *Aloe chabaudii* Schönland var. *chabaudii*
Pretoria aloe . . . . . . *Aloe pretoriensis* Pole-Evans
Pretoria-aalwyn . . . . . . *Aloe pretoriensis* Pole-Evans
pulpos . . . . . . *Aloe arborescens* Mill. subsp. *arborescens*
purple spotted aloe . . . . . . *Aloe parvibracteata* Schönland

quiver tree . . . . . . *Aloe dichotoma* Masson

raktapolam . . . . . . *Aloe littoralis* Baker
ramadanhi . . . . . . *Aloe buettneri* A.Berger

ramenas	*Aloe longistyla* Baker
ranga	*Aloe schweinfurthii* Baker
rangambala	*Aloe schweinfurthii* Baker
rangambia	*Aloe schweinfurthii* Baker
rankaalwee	*Aloe gracilis* Haw.
rankaalwyn	*Aloe decumbens* (Reynolds) van Jaarsv.
	Aloe gracilis Haw.
rat aloe	*Aloe ballyi* Reynolds
rauhblättrige aloe	*Aloe asperifolia* A.Berger
're harmaz	*Aloe percrassa* Tod.
're	*Aloe* L.
red aloe	*Aloe ferox* Mill.
red aloes	*Aloe ferox* Mill.
red hot poker aloe	*Aloe aculeata* Pole-Evans
red spined striped aloe	*Aloe lineata* (Aiton) Haw. var. *lineata*
red-hot poker	*Aloe peglerae* Schönland
reetii	*Aloe trichosantha* A.Berger subsp. *trichosantha*
regte-aalwyn	*Aloe ferox* Mill.
	Aloe marlothii A.Berger subsp. *marlothii*
Reitz's aloe	*Aloe reitzii* Reynolds var. *reitzii*
reusekokerboom	*Aloe pillansii* L.Guthrie
Rhodesian aloe	*Aloe rhodesiana* Rendle
Rhodesian tree aloe	*Aloe excelsa* A.Berger var. *excelsa*
`riet	*Aloe* L.
rock aloe	*Aloe petricola* Pole-Evans
	Aloe rupestris Baker
rokaalwyn	*Aloe alooides* (Bolus) Druten
rooi-aalwyn	*Aloe gariepensis* Pillans
rora	*Aloe ballyi* Reynolds
rotsaalwyn	*Aloe petricola* Pole-Evans
rukaka	*Aloe* L.
rumangamunu	*Aloe excelsa* A.Berger var. *excelsa*
rumangamuru	*Aloe excelsa* A.Berger var. *excelsa*
rumhangamhuno	*Aloe* L.
	Aloe chabaudii Schönland var. *chabaudii*
	Aloe excelsa A.Berger var. *excelsa*
	Aloe greatheadii Schönland
ruvati	*Aloe* L.
	Aloe chabaudii Schönland var. *chabaudii*
	Aloe excelsa A.Berger var. *excelsa*
	Aloe greatheadii Schönland

saber socotri . . . . *Aloe perryi* Baker
sábila china . . . . *Aloe ciliaris* Haw. var. *ciliaris*
sábila do penca . . . . *Aloe vera* (L.) Burm.f.
sabila . . . . *Aloe arborescens* Mill. subsp. *arborescens*
Aloe succotrina Weston
Aloe vera (L.) Burm.f.
sábila . . . . *Aloe vera* (L.) Burm.f.
sabila-pinya . . . . *Aloe vera* (L.) Burm.f.
sabir suqutri . . . . *Aloe perryi* Baker
sabir . . . . *Aloe succotrina* Weston
sahondra . . . . *Aloe capitata* Baker var. *capitata*
sakoankenke . . . . *Aloe massawana* Reynolds subsp. *sakoankenke* (J.-B.Castillon) J.-B.Castillon
sand aalwyn . . . . *Aloe hereroensis* Engl. var. *hereroensis*
sand aloe . . . . *Aloe hereroensis* Engl. var. *hereroensis*
sanda `re . . . . *Aloe camperi* Schweinf.
sandaalwyn . . . . *Aloe hereroensis* Engl. var. *hereroensis*
sankulu . . . . *Aloe zebrina* Baker
saqal . . . . *Aloe vera* (L.) Burm.f.
sarısabır . . . . *Aloe vera* (L.) Burm.f.
sarivahona . . . . *Aloe haworthioides* Baker var. *aurantiaca* H.Perrier
sarýsabýr . . . . *Aloe vera* (L.) Burm.f.
sasparila . . . . *Aloe vera* (L.) Burm.f.
sávila . . . . *Aloe vera* (L.) Burm.f.
sawila . . . . *Aloe vera* (L.) Burm.f.
sawupo . . . . *Aloe castanea* Schönland
sayyan . . . . *Aloe arborescens* Mill. subsp. *arborescens*
scrambling aloe . . . . *Aloe gracilis* Haw.
sea-side aloe . . . . *Aloe littoralis* Baker
seepaalwyn . . . . *Aloe maculata* All.
seholobe . . . . *Aloe striatula* Haw. var. *striatula*
sekgopha . . . . *Aloe greatheadii* Schönland
sekope . . . . *Aloe aculeata* Pole-Evans
senjela . . . . *Aloe* L.
senjere . . . . *Aloe menyharthii* Baker subsp. *menyharthii*
senjerere . . . . *Aloe* L.
sereberebe . . . . *Aloe buettneri* A.Berger
sere-berebe . . . . *Aloe buettneri* A.Berger
serelei . . . . *Aloe aristata* Haw.
Aloe humilis (L.) Mill.
sereleli . . . . *Aloe aristata* Haw.
Aloe ecklonis Salm-Dyck

Sesfonteinaalwyn . . . . . *Aloe dewinteri* Giess
set `ret . . . . . *Aloe pulcherrima* M.G.Gilbert & Sebsebe
short-leaved aloe. . . . . *Aloe distans* Haw.
sibr . . . . . *Aloe succotrina* Weston
sikorowet . . . . . *Aloe kedongensis* Reynolds
Aloe lateritia Engl. var. *lateritia*
Aloe secundiflora Engl. var. *secundiflora*
silver-tailed aloe . . . . . *Aloe argenticauda* Merxm. & Giess
siniani yebo . . . . . *Aloe macrocarpa* Tod. subsp. *macrocarpa*
Aloe maculata All.
sinzé toro . . . . . *Aloe buettneri* A.Berger
siroo-luttalay . . . . . *Aloe littoralis* Baker
siru karalai . . . . . *Aloe littoralis* Baker
sizimyamuliro . . . . . *Aloe lateritia* Engl. var. *lateritia*
skirt aloe . . . . . *Aloe alooides* (Bolus) Druten
slangaalwyn . . . . . *Aloe broomii* Schönland var. *broomii*
slangkop . . . . . *Aloe pruinosa* Reynolds
slangkopaalwyn . . . . . *Aloe pruinosa* Reynolds
slapare-aalwyn . . . . . *Aloe speciosa* Baker
slaphoringaalwyn . . . . . *Aloe speciosa* Baker
slapoor . . . . . *Aloe speciosa* Baker
slapooraalwyn . . . . . *Aloe speciosa* Baker
snuifaalwyn . . . . . *Aloe marlothii* A.Berger subsp. *marlothii*
soap aloe. . . . . *Aloe maculata* All.
Socotra aloe . . . . . *Aloe perryi* Baker
Socotrine aloe. . . . . *Aloe perryi* Baker
Socotrine aloes . . . . . *Aloe succotrina* Weston
Socotrine du pays . . . . . *Aloe purpurea* Lam.
soft distant sword-leaved aloe. . . . . *Aloe gracilicaulis* Reynolds & P.R.O.Bally
sogoba bu . . . . . *Aloe buettneri* A.Berger
sogoba hu . . . . . *Aloe buettneri* A.Berger
sogoba ku . . . . . *Aloe buettneri* A.Berger
spaansaalwee. . . . . *Aloe speciosa* Baker
Aloe spicata L.f.
spaansaalwyn. . . . . *Aloe speciosa* Baker
Aloe spicata L.f.
spanareaalwyn . . . . . *Aloe speciosa* Baker
spanarei-aalwee . . . . . *Aloe speciosa* Baker
Aloe spicata L.f.
spanarei-aalwyn . . . . . *Aloe speciosa* Baker
Aloe spicata L.f.

spansaalwyn . . . . . *Aloe speciosa* Baker
spiraalaalwyn . . . . . *Aloe polyphylla* Schönland ex Pillans
spiral aloe . . . . . *Aloe polyphylla* Schönland ex Pillans
spire aloe . . . . . *Aloe cryptopoda* Baker
spotted aloe hybrid . . . . . *Aloe* ×*schimperi* Tod.
spotted aloe . . . . . *Aloe affinis* A.Berger
Aloe davyana Schönland
Aloe parvibracteata Schönland
Aloe prinslooi I.Verd. & D.S.Hardy
Aloe zebrina Baker
star cactus . . . . . *Aloe vera* (L.) Burm.f.
strand aloe . . . . . *Aloe thraskii* Baker
strandaalwyn . . . . . *Aloe distans* Haw.
Aloe thraskii Baker
strandveldaalwyn . . . . . *Aloe arenicola* Reynolds
Aloe thraskii Baker
streepaalwyn . . . . . *Aloe striata* Haw.
stripe-sheathed narrow-leaved aloe . . . . . *Aloe striatula* Haw. var. *striatula*
struikaalwyn . . . . . *Aloe pearsonii* Schönland
subiri . . . . . *Aloe lateritia* Engl. var. *lateritia*
suk' ro-i . . . . . *Aloe secundiflora* Engl. var. *secundiflora*
sukoroi . . . . . *Aloe secundiflora* Engl. var. *secundiflora*
suwopa . . . . . *Aloe castanea* Schönland
swartdoringaalwyn . . . . . *Aloe melanacantha* A.Berger
Swellendam-aalwee . . . . . *Aloe ferox* Mill.
Swellendamaalwyn . . . . . *Aloe ferox* Mill.
Swellendamsaalwee . . . . . *Aloe ferox* Mill.
Swellendamsaalwyn . . . . . *Aloe ferox* Mill.
sword aloe . . . . . *Aloe arborescens* Mill. subsp. *arborescens*
Swynnerton's aloe . . . . . *Aloe swynnertonii* Rendle
Swynnerton's spotted leaf aloe . . . . . *Aloe swynnertonii* Rendle
sybir . . . . . *Aloe littoralis* Baker

Table Mountain aloe . . . . . *Aloe succotrina* Weston
Tafelbergaalwyn . . . . . *Aloe succotrina* Weston
tangaratuet . . . . . *Aloe kedongensis* Reynolds
Aloe lateritia Engl. var. *lateritia*
Aloe secundiflora Engl. var. *secundiflora*
tangaratwe . . . . . *Aloe dawei* A.Berger
tangaratwet . . . . . *Aloe dawei* A.Berger
Aloe kedongensis Reynolds

tangaratwet (cont.)	*Aloe lateritia* Engl. var. *lateritia*
	Aloe secundiflora Engl. var. *secundiflora*
tap aloe	*Aloe ferox* Mill.
tapaalwee	*Aloe ferox* Mill.
tapaalwyn	*Aloe ferox* Mill.
tap-aalwyn	*Aloe ferox* Mill.
	Aloe marlothii A.Berger subsp. *marlothii*
tarentaalaalwyn	*Aloe aristata* Haw.
tati	*Aloe maculata* All.
tayf za'alhil	*Aloe perryi* Baker
tembwisya	*Aloe nuttii* Baker
Thompson's aloe	*Aloe thompsoniae* Groenew.
thugurui	*Aloe kedongensis* Reynolds
	Aloe lateritia Engl. var. *lateritia*
	Aloe secundiflora Engl. var. *secundiflora*
tienkara sansugu	*Aloe buettneri* A.Berger
tiger aloe	*Aloe variegata* L.
	Aloe zebrina Baker
tilt-head aloe	*Aloe speciosa* Baker
toba xa	*Aloe vera* (L.) Burm.f.
toba xha	*Aloe vera* (L.) Burm.f.
tolkos	*Aloe kedongensis* Reynolds
	Aloe lateritia Engl. var. *graminicola* (Reynolds) S.Carter
	Aloe lateritia Engl. var. *lateritia*
	Aloe secundiflora Engl. var. *secundiflora*
tongaalwyn	*Aloe plicatilis* (L.) Mill.
tookgo	*Aloe fosteri* Pillans
toots amarelo	*Aloe vera* (L.) Burm.f.
torch plant	*Aloe arborescens* Mill. subsp. *arborescens*
torchplant	*Aloe arborescens* Mill. subsp. *arborescens*
torch-plant	*Aloe aristata* Haw.
Transvaal aalwyn	*Aloe davyana* Schönland
Transvaal aloe	*Aloe marlothii* A.Berger subsp. *marlothii*
Transvaalaalwyn	*Aloe davyana* Schönland
tree aloe	*Aloe barberae* Dyer
	Aloe excelsa A.Berger var. *excelsa*
	Aloe marlothii A.Berger subsp. *marlothii*
true aloe	*Aloe vera* (L.) Burm.f.
tshikhopa	*Aloe micracantha* Haw.
tshikhopha	*Aloe arborescens* Mill. subsp. *arborescens*
	Aloe chabaudii Schönland var. *chabaudii*
	Aloe excelsa A.Berger var. *excelsa*

tshikhopha (cont.) . . . . *Aloe littoralis* Baker
Aloe marlothii A.Berger subsp. *marlothii*
Aloe spicata L.f.

Tshipise-aalwyn . . . . *Aloe lutescens* Groenew.

tshiṱuku . . . . *Aloe micracantha* Haw.

tsikivahonbaho . . . . *Aloe antandroi* (Decary) H.Perrier subsp. *antandroi*

tsikyvahombaho . . . . *Aloe antandroi* (Decary) H.Perrier subsp. *antandroi*

tulédagbla . . . . *Aloe buettneri* A.Berger

Turk's cap aloe . . . . *Aloe peglerae* Schönland

Turkey aloe . . . . *Aloe succotrina* Weston

tweederly aloëboom . . . . *Aloe khamiesensis* Pillans

tweederly . . . . *Aloe khamiesensis* Pillans

tyf za'abeb . . . . *Aloe perryi* Baker

Uitenhaags-aalwee . . . . *Aloe africana* Mill.

Uitenhaagsaalwyn . . . . *Aloe africana* Mill.

Uitenhaagse aalwyn . . . . *Aloe africana* Mill.
Aloe striata Haw.

Uitenhaagse-aalwee . . . . *Aloe africana* Mill.

Uitenhaagseaalwyn . . . . *Aloe africana* Mill.

Uitenhaagse-aalwyn . . . . *Aloe africana* Mill.

Uitenhage aloe . . . . *Aloe africana* Mill.

uluphondonde . . . . *Aloe rupestris* Baker

umakhuphulwane . . . . *Aloe myriacantha* (Haw.) Schult. & Schult.f.

umathithibala . . . . *Aloe aristata* Haw.

umgxwala . . . . *Aloe barberae* Dyer

umgzwala . . . . *Aloe barberae* Dyer

umhlaba . . . . *Aloe candelabrum* A.Berger
Aloe ferox Mill.
Aloe marlothii A.Berger subsp. *marlothii*
Aloe spectabilis Reynolds
Aloe spicata L.f.
Aloe thraskii Baker

um-hlaba . . . . *Aloe marlothii* A.Berger subsp. *marlothii*

umhlabana . . . . *Aloe arborescens* Mill. subsp. *arborescens*

umhlabandlanzi . . . . *Aloe barberae* Dyer
Aloe rupestris Baker

umhlabandlazi . . . . *Aloe rupestris* Baker

umhlabanhazi. . . . . *Aloe rupestris* Baker

umhlabanhlazi . . . . *Aloe rupestris* Baker
Aloe suprafoliata Pole-Evans

umhlakahla . . . . *Aloe ferox* Mill.
umhlalampofu . . . . *Aloe barberae* Dyer
umjinqa . . . . *Aloe tenuior* Haw.
umkala . . . . *Aloe marlothii* A.Berger subsp. *marlothii*
Aloe spectabilis Reynolds
umpondonde . . . . *Aloe barberae* Dyer
Aloe rupestris Baker
undlampofu . . . . *Aloe rupestris* Baker
undyang'oma . . . . *Aloe menyharthii* Baker subsp. *menyharthii*
unguentine cactus . . . . *Aloe vera* (L.) Burm.f.
unomaweni . . . . *Aloe arborescens* Mill. subsp. *arborescens*
Aloe ferox Mill.
unqcelwane . . . . *Aloe striatula* Haw. var. *striatula*
uolaganti . . . . *Aloe pirottae* A.Berger
uphondonde . . . . *Aloe rupestris* Baker
upondonde . . . . *Aloe rupestris* Baker
uwindi . . . . *Aloe menyharthii* Baker subsp. *menyharthii*

vaalblaaraalwyn . . . . *Aloe striata* Haw.
vaalblaar-aalwyn . . . . *Aloe striata* Haw.
vaalvygie . . . . *Aloe melanacantha* A.Berger
vaho . . . . *Aloe aurelienii* J.-B.Castillon
Aloe teissieri Lavranos
vahomafaitra . . . . *Aloe divaricata* A.Berger var. *divaricata*
vahombato . . . . *Aloe deltoideodonta* Baker var. *deltoideodonta*
Aloe imalotensis Reynolds var. *imalotensis*
vahombe . . . . *Aloe macroclada* Baker
Aloe vaombe Decorse & Poiss. var. *vaombe*
vahona . . . . *Aloe divaricata* A.Berger var. *divaricata*
Aloe macroclada Baker
vahondrano . . . . *Aloe suzannae* Decary
vahongarana . . . . *Aloe deltoideodonta* Baker var. *deltoideodonta*
Aloe imalotensis Reynolds var. *imalotensis*
vahonmafaitra . . . . *Aloe divaricata* A.Berger var. *divaricata*
vahonomafaitra . . . . *Aloe divaricata* A.Berger var. *divaricata*
vahonona . . . . *Aloe cipolinicola* (H.Perrier) J.-B.Castillon & J.-P.Castillon
vahontsohy . . . . *Aloe divaricata* A.Berger var. *divaricata*
vahotsanda . . . . *Aloe vaotsanda* Decary
vahotsohy . . . . *Aloe divaricata* A.Berger var. *divaricata*
Van Balen's aloe . . . . *Aloe vanbalenii* Pillans
Vanrhynsdorpaalwyn . . . . *Aloe falcata* Baker

variegated aloe	*Aloe variegata* L.
vela	*Aloe arborescens* Mill. subsp. *arborescens*
Vereker's dwala aloe	*Aloe chabaudii* Schönland var. *chabaudii*
very much branched aloe	*Aloe ramosissima* Pillans
vlakteaalwyn	*Aloe hereroensis* Engl. var. *hereroensis*
vlakte-aalwyn	*Aloe hereroensis* Engl. var. *hereroensis*
vlei-aalwyn	*Aloe ecklonis* Salm-Dyck
vohandranjo	*Aloe divaricata* A.Berger var. *divaricata*
Vryheid aloe	*Aloe vryheidensis* Groenew.
vuurpylaalwyn	*Aloe peglerae* Schönland
waaieraalwyn	*Aloe plicatilis* (L.) Mill.
waaier-aalwyn	*Aloe plicatilis* (L.) Mill.
wantchile	*Aloe christianii* Reynolds
wateraalwee	*Aloe micracantha* Haw.
wateraalwyn	*Aloe micracantha* Haw.
wend `riet	*Aloe trichosantha* A.Berger subsp. *trichosantha*
West African aloe	*Aloe buettneri* A.Berger
West Indian aloe	*Aloe vera* (L.) Burm.f.
white spotted aloe	*Aloe maculata* All.
white-thorn aloe	*Aloe aculeata* Pole-Evans
Wickens' aloe	*Aloe wickensii* Pole-Evans var. *wickensii*
Wickens-aalwyn	*Aloe wickensii* Pole-Evans var. *wickensii*
Wild's small Chimanimani aloe	*Aloe wildii* (Reynolds) Reynolds
wilde-aalwee	*Aloe khamiesensis* Pillans
wilde-aalwyn	*Aloe khamiesensis* Pillans
wildeaalwyn	*Aloe khamiesensis* Pillans *Aloe speciosa* Baker *Aloe spicata* L.f.
Windhoek aloe	*Aloe littoralis* Baker
Windhoekaalwyn	*Aloe littoralis* Baker
witchdoctor's aloe	*Aloe globuligemma* Pole-Evans
witdoringaalwyn	*Aloe aculeata* Pole-Evans
Wolkberg aloe	*Aloe vryheidensis* Groenew.
Wolkbergaalwyn	*Aloe vryheidensis* Groenew.
woody aloe	*Aloe arborescens* Mill. subsp. *arborescens*
wudie	*Aloe buettneri* A.Berger
wundumba	*Aloe excelsa* A.Berger var. *excelsa*
Wylliespoort aloe	*Aloe angelica* Pole-Evans
Wylliespoortaalwyn	*Aloe angelica* Pole-Evans

xiteti . . . . *Aloe barberae* Dyer
xitretre . . . . *Aloe arborescens* Mill. subsp. *arborescens*
Aloe parvibracteata Schönland

yalva . . . . *Aloe succotrina* Weston
yeliyo . . . . *Aloe succotrina* Weston
yellow aloe . . . . *Aloe cryptopoda* Baker
yellow spineless aloe . . . . *Aloe reynoldsii* Letty
yerba babose . . . . *Aloe succotrina* Weston
yerba del acibar . . . . *Aloe succotrina* Weston

zààbóó . . . . *Aloe buettneri* A.Berger
zaabuwaa . . . . *Aloe buettneri* A.Berger
zaber . . . . *Aloe percrassa* Tod.
zábila dos toots . . . . *Aloe vera* (L.) Burm.f.
zabila . . . . *Aloe arborescens* Mill. subsp. *arborescens*
Aloe maculata All.
Aloe vera (L.) Mill.
zábila . . . . *Aloe vera* (L.) Burm.f.
zabo . . . . *Aloe buettneri* A.Berger
zaboko . . . . *Aloe buettneri* A.Berger
zabon dafi . . . . *Aloe buettneri* A.Berger
zabur . . . . *Aloe* L.
zabuwa . . . . *Aloe buettneri* A.Berger
zadiva . . . . *Aloe vera* (L.) Burm.f.
Zambezi gorges aloe . . . . *Aloe chabaudii* Schönland var. *chabaudii*
zanana mowo . . . . *Aloe aldabrensis* (Marais) L.E.Newton & G.D.Rowley
Zanzibar aloe . . . . *Aloe perryi* Baker
Aloe succotrina Weston
zapua disum . . . . *Aloe buettneri* A.Berger
zebra leaf aloe . . . . *Aloe zebrina* Baker
zebra . . . . *Aloe arborescens* Mill. subsp. *arborescens*
zebra-leaf aloe . . . . *Aloe zebrina* Baker
Zimbabwe aloe . . . . *Aloe excelsa* A.Berger var. *excelsa*
Zimbabwe tree aloe . . . . *Aloe excelsa* A.Berger var. *excelsa*
Zimbabwe-aalwyn . . . . *Aloe excelsa* A.Berger var. *excelsa*
zotollin . . . . *Aloe vera* (L.) Burm.f.

Алое настоящее (aloe nastojaščee) . . . . *Aloe vera* (L.) Burm.f.
Алоэ (aloe) . . . . *Aloe vera* (L.) Burm.f.

Алоэ Вера (aloe vera) . . . . *Aloe vera* (L.) Burm.f.
الألوة, الألوه نبات. . . . . *Aloe vera* (L.) Burm.f.
وبراد . . . . *Aloe vera* (L.) Burm.f.
ว่านไฟไหม้ (wan fai mai) . . . . *Aloe vera* (L.) Burm.f.
ว่านหางจระเข้ (wan hang chora khe) . . . . *Aloe vera* (L.) Burm.f.
หางตะเข้ (hang ta khe). . . . . *Aloe vera* (L.) Burm.f.
ア ロ エ (aroe) . . . . *Aloe vera* (L.) Burm.f.
ア ロ エ チ ヨ ダ ニ シ キ (aroe chiyodanishiki). . . . . *Aloe zebrina* Baker
龙舌兰. . . . . *Aloe vera* (L.) Burm.f.
千代田錦 . . . . *Aloe zebrina* Baker

Aloe khamiesensis (Photographer: A.W. Klopper)

Products

Product	Source species
Aloe	*Aloe ferox* Mill. *Aloe vera* (L.) Burm.f.
Aloes	*Aloe* L.
Aloe Barbadensis	*Aloe vera* (L.) Burm.f.
Aloe Capensis	*Aloe ferox* Mill.
Aloe Vera Gel	*Aloe vera* (L.) Burm.f.
Barbados aloes	*Aloe vera* (L.) Burm.f.
Cape aloes	*Aloe ferox* Mill.
Curaçao aloes	*Aloe vera* (L.) Burm.f.
Drug aloes	*Aloe* L.
Natal aloes	*Aloe spectabilis* Reynolds
Powdered aloes	*Aloe* L.
Socotrine aloes	*Aloe perryi* Baker
Zanzibar aloes	*Aloe* L.

Source species

Source species	Product
Aloe L.	Aloes Drug aloes Powdered aloes Zanzibar aloes
Aloe ferox Mill.	Aloe Aloe Capensis Cape aloes
Aloe perryi Baker	Socotrine aloes
Aloe spectabilis Reynolds	Natal aloes
Aloe vera (L.) Burm.f.	Aloe Aloe Barbadensis Aloe Vera Gel Barbados aloes Curaçao aloes

STRELITZIA

1. Botanical diversity in southern Africa. 1994. B.J. Huntley (ed.). ISBN 1-874907-25-0.
2. Cyperaceae in Natal. 1995. K.D. Gordon-Gray. ISBN 1-874907-04-8.
3. Cederberg vegetation and flora. 1996. H.C. Taylor. ISBN 1-874907-28-5.
4. Red Data List of southern African plants. 1996. Craig Hilton-Taylor. ISBN 1-874907-29-3.
5. Taxonomic literature of southern African plants. 1997. N.L. Meyer, M. Mössmer & G.F. Smith (eds). ISBN 1-874907-35-8.
6. Plants of the northern provinces of South Africa: keys and diagnostic characters. 1997. E. Retief & P.P.J. Herman. ISBN 1-874907-30-7.
7. Preparing herbarium specimens. 1999. Lyn Fish. ISBN 1-919795-38-3.
8. Bulbinella in South Africa. 1999. Pauline L. Perry. ISBN 1-919795-46-4. OUT OF PRINT.
9. Cape plants. A conspectus of the Cape flora of South Africa. 2000. P. Goldblatt & J.C. Manning. ISBN 0-620-26236-2.
10. Seed plants of southern Africa: families and genera. 2000. O.A. Leistner (ed.). ISBN 1-919795-51-0.
11. The Cape genus Lachnaea (Thymelaeaceae): a monograph. 2001. J.B.P. Beyers. ISBN 1-919795-52-9.
12. The Global Taxonomy Initiative: documenting the biodiversity of Africa/L'Initiative Taxonomique Mondiale: documenter la biodiversité en Afrique. R.R. Klopper, G.F. Smith & A.C. Chikuni (eds). 2001. ISBN 1-919795-63-4. OUT OF PRINT.
13. Medicinal and magical plants of southern Africa: an annotated checklist. 2002. T.H. Arnold, C.A. Prentice, L.C. Hawker, E.E. Snyman, M. Tomalin, N.R. Crouch & C. Pottas-Bircher. ISBN 1-919795-62-6.
14. Plants of southern Africa: an annotated checklist. 2003. G. Germishuizen & N.L. Meyer (eds). ISBN 1-919795-99-5.
15. Heyday of the gymnosperms: systematics and biodiversity of the Late Triassic Molteno fructifications. 2003. J.M. Anderson & H.M. Anderson. ISBN 1-919795-98-7.
16. Common names of Karoo plants. 2004. Les Powrie. ISBN 1-874907-16-1.
17. National Spatial Biodiversity Assessment 2004: priorities for biodiversity conservation in South Africa. 2005. A. Driver, K. Maze, M. Rouget, A.T. Lombard, J. Nel, J.K. Turpie, R.M. Cowling, P. Desmet, P. Goodman, J. Harris, Z. Jonas, B. Reyers, K. Sink & T. Strauss. ISBN 1-919976-20-5.
18. A revision of the southern African genus Babiana, Iridaceae: Crocoideae. 2007. P. Goldblatt & J.C. Manning. ISBN-10: 1-919976-32-9. ISBN-13: 978-1-919976-32-7.
19. The vegetation of South Africa, Lesotho and Swaziland. 2006. L. Mucina & M.C. Rutherford (eds). ISBN-10: 1-919976-21-3. ISBN-13: 978-1-919976-21-1.
20. Brief history of the gymnosperms: classification, biodiversity, phytogeography and ecology. 2007. J.M. Anderson, H.M. Anderson & C.J. Cleal. ISBN 978-1-919976-39-6.
21. Molteno ferns: Late Triassic biodiversity in southern Africa. 2008. H.M. Anderson & J.M. Anderson. ISBN 978-1-919976-36-5.
22. Plants of Angola / Plantas de Angola. 2008. E. Figueiredo & G.F. Smith. ISBN 978-1-919976-45-7.
23. Synopsis of the Lycopodiophyta and Pteridophyta of Africa, Madagascar and neighbouring islands. 2009. J.P. Roux. ISBN 978-1-919976-48-8.
24. Historical plant incidence in southern Africa. 2009. C.J. Skead. ISBN 978-1-919976-53-2.
25. Red List of South African plants 2009. 2009. D. Raimondo, L. von Staden, W. Foden, J.E. Victor, N.A. Helme, R.C. Turner, D.A. Kamundi & P.A. Manyama (eds). ISBN 978-1-919976-52-5.
26. Botanical exploration of southern Africa, edn 2. 2010. H.F. Glen & G. Germishuizen. ISBN 978-1-919976-54-9.
27. Botany and horticulture of the genus Freesia (Iridaceae). 2010. J.C. Manning & P. Goldblatt (with G.D. Duncan, F. Forest, R. Kaiser & L. Tatarenko). Paintings by Auriol Batten; line drawings by John C. Manning. ISBN 978-1-919976-58-7.
28 The aloe names book. 2011. O.M. Grace, R.R. Klopper, E. Figueiredo & G.F. Smith. ISBN 978-1-919976-64-8

ENQUIRIES:

Bookshop, South African National Biodiversity Institute, Private Bag X101, Pretoria, 0001 South Africa.
Tel. +27 12 843-5000 • Fax +27 12 804-3211
E-mail bookshop@sanbi.org.za • http://www.sanbi.org